THE DEVINE EQUATION

LANCE OVERMAN

The Devine Equation

By Lance Overman

Copyright © Lance Overman

All rights reserved. No part of this book may be reproduced, distributed, or transmitted in any form or by any means, including photocopying, recording, or other electronic or mechanical methods, without the prior written permission of the publisher, except in the case of brief quotations embodied in critical reviews and certain other noncommercial uses permitted by copyright law. For permission requests, write to the publisher

Published by
Lance Overman Publishing

First Edition: September 2024
ISBN: 9798342695695
This is a work of fiction. Names, characters, businesses, places, events, and incidents are either the products of the author's imagination or used in a fictitious manner. Any resemblance to actual persons, living or dead, or actual events is purely coincidental.

Thank you for reading and supporting the author's work.

Dedication

To the brilliant mathematicians who have dared to peer beyond the veil of the ordinary, who have uncovered the hidden symmetries and elegant patterns that govern our universe, this work is humbly dedicated.

Your unyielding pursuit of truth, your capacity to see the world in numbers and abstract forms, has not only unlocked the mysteries of existence but has also laid the foundation for others to find meaning in the complexity of life. Through your equations and theories, we glimpse the infinite–fractals of possibility that inspire us to seek deeper connections between the tangible and the transcendent.

In gratitude for the beauty you have shown us, and for the countless ways in which your work continues to illuminate the path ahead.

Theory of Everything

$$W = \int_{k<\Lambda} [Dg][DA][D\psi][D\Phi] \exp\left\{ i \int d^4x \sqrt{-g} \left[\frac{m_p^2}{2} R \right.\right.$$

$$\left.\left. -\frac{1}{4} F^a_{\mu\nu} F^{a\mu\nu} + i\bar{\psi}^i \gamma^\mu D_\mu \psi^i + \left(\bar{\psi}^i_L V_{ij} \Phi \psi^j_R + \text{h.c.} \right) - |D_\mu \Phi|^2 - V(\Phi) \right] \right\}$$

quantum mechanics — spacetime gravity

other forces — matter — Higgs

1. $\int$ denotes integration.

2. $k < \Lambda$ represents the integration limits or conditions.

3. $[Dg][DA][D\psi][D\Phi]$ represents path integrals over fields g, A, ψ, and Φ.

4. exp refers to the exponential function.

5. i is the imaginary unit.

6. d^4x represents a four-dimensional space-time integration.

7. $\sqrt{(-g)}$ is the square root of the determinant of the metric tensor (in relativity).

8. $\frac{m_p^2}{2} R$ represents a term involving the Ricci scalar R, related to gravity.

9. $-\frac{1}{4} F^a_{\mu\nu} F^{a\mu\nu}$ represents the field strength tensor squared, commonly appearing in gauge theories.

10. $i\bar{\psi}^i \gamma^\mu D_\mu \psi^i$ involves Dirac spinors and gamma matrices.

11. $(\bar{\psi}L^i V_{ij} \Phi \psi^j_R + h.c.)$ represents interaction terms, including Yukawa interactions with the Higgs field.

12. $|D_\mu \Phi|^2$ and $V(\Phi)$ are terms involving the Higgs field and its potential.

PROLOGUE

In the beginning, there was chaos. A primal stew of infinite possibilities, swirling in the void, formless yet brimming with potential. The universe itself was a kitchen, and the forces within it–the cosmic ingredients–awaited the hands of a master chef to craft them into something meaningful, something sublime.
Jaxon Keller was born into a world that seemed orderly on the surface. A world of sharp knives and pristine kitchens, of exact measurements and predictable results. He flourished in this world, his talents propelling him to the heights of culinary fame, where precision was worshiped and perfection was the only acceptable outcome. Yet, beneath the gleaming surfaces and the accolades, a storm brewed–unseen, insidious, unstoppable.
It began as a whisper–a subtle dissonance between the external order he imposed on the world and the internal chaos that roiled within him. A chef's mind, after all, is a delicate balance of creativity and control, of passion and discipline. But when that balance tips, when the harmony between order and chaos is lost, the result can be catastrophic.
Jaxon's fall was swift and brutal, his mind unraveling in a blur of addiction and despair. The kitchen, once his sanctuary, became a battlefield where he waged war against himself. The ingredients that once spoke to him in a language of flavors and textures now mocked him, their secrets locked away behind a haze of alcohol and drugs.
It was in this abyss that the lightning found him, searing his soul with a revelation that would forever change the trajectory of his life. In that blinding flash, Jaxon glimpsed the truth that had eluded him for so long–that the chaos he feared was not his enemy, but his greatest ally. The universe, in all its complexity and wonder, was a fractal–a pattern of infinite beauty and order, hidden within the very chaos he had sought to escape.
And so, Jaxon emerged from the storm, reborn as both a chef and a seeker. His journey was no longer just about creating exquisite dishes; it was about understanding the cosmic recipe that governed all existence. He became an alchemist of sorts, blending the culinary with the cosmic, using food as a medium to explore the deepest mysteries of the universe.
This is his story–a story of redemption and revelation, of the delicate dance between order and chaos, and of the infinite fractals that connect us all.

Chapter 1

THE EQUATION OF ALL THINGS

Jaxon awoke to the pungent scent of decay, a stench that seemed to curl around the edges of reality like a sinister Möbius strip, unending and inescapable. His apartment was a monochrome tangle of shadows, the dim light filtering through the dirty blinds casting angular patterns across the stained couch he was sprawled upon. He blinked slowly, his consciousness rising like a tangled equation struggling toward solvability. The remnants of last night's indulgence–the empty bottles and scattered paraphernalia–were strewn across the floor, their chaotic arrangement mimicking the random distribution of particles in an entropic system.

His body felt sluggish, as though his limbs were composed of heavy, viscous matter rather than flesh and bone. Each attempt to move was met with the resistance of inertia, a law of physics manifesting in the very fibers of his being. The hollow clatter of bottles hitting the floor reverberated through the room, a stark reminder of the void within him. It was as if he were the focal point of a collapsing star, his essence imploding under the weight of addiction's relentless gravity.

Dragging himself to his feet, he moved toward the kitchen, a space that had once been his domain–a laboratory of culinary innovation where every dish was a complex algorithm of flavors, calculated and refined to perfection. Now, it was a wasteland, a physical manifestation of the second law of thermodynamics, where disorder only increased with time. The sink, overflowing with dishes crusted in dried remnants of forgotten meals, resembled a festering Petri dish, breeding bacteria and decay. Fast food wrappers and rotting takeout containers were scattered about like abandoned variables in a failed proof, each one a testament to the neglect that mirrored his internal desolation.

The air was thick with a miasma of rot and forgotten dreams, each breath a struggle against the atmosphere of hopelessness. Jaxon's once-keen sense of smell, the instrument that had guided his culinary genius, was now dulled by the stench of his surroundings—a fitting metaphor for the decay of his once-sharp mind. The floor beneath his feet, sticky with spilled liquor and crumbs, was a crude map of his recent past, each step a journey through the wasteland of his self-destruction.

Rummaging through the debris, Jaxon found nothing of sustenance, nothing to nourish the void that consumed him from within. Every discarded wrapper, every half-eaten meal, was a symbol of the dreams he had allowed to rot, the ambitions he had systematically dismantled through the numbing haze of intoxication. The tangible neglect of his environment was a fractal of his own psyche, endlessly repeating patterns of decay and hopelessness at every scale of his existence.

As he stood there, surveying the wreckage of his life, his mind began to spiral into the labyrinth of his past. He could almost see the golden ratio in the curve of a perfectly plated dish, the Fibonacci sequence in the arrangement of ingredients on a cutting board. But those patterns, once as clear as a Euclidean proof, were now lost to him, buried under the noise of addiction. The kitchen, once a temple to his art, had become a tomb, and Jaxon its unwilling occupant.

His phone buzzed on the counter, a brief flicker of light in the encroaching darkness. The messages that awaited him were a series of lifelines, threads of concern sent by friends who were becoming increasingly distant in the space-time continuum of his life. He stared at the screen, the words blurring together as if viewed through a gravitational lens, distorted by the mass of his despair. Their messages were filled with concern, saturated with frustration, and laced with a love that felt as alien to him as quantum mechanics to a layman. They begged him to reconnect, to seek help, to return to the man they once knew. But to Jaxon, these pleas were like echoes bouncing off the event horizon of a black hole—forever out of reach, impossible to grasp.

He cast the phone aside, the device clattering to the floor with a sound that seemed to reverberate through his very soul. Stumbling to a worn-out chair, he collapsed in front of the television, his mind numbly absorbing the images of culinary mastery that flickered across the screen. Cooking shows paraded before his eyes, each one a cruel reminder of the craft he had once commanded with the precision of a physicist manipulating the forces of the universe. The chefs on the screen danced with their ingredients, their movements an elegant ballet of knife and fire, transforming raw elements into masterpieces of taste and texture.

In these moments, Jaxon felt the pangs of a distant pride, a pulse of life that still beat somewhere deep within him, buried under layers of addiction and despair. His memories drifted back to his days of glory—the rhythmic dance of knives, the sizzle of pans, the delicate interplay of flavors and textures that had once defined his very existence. He had been a conductor in a symphony of taste, each note meticulously crafted to resonate with the palates of those fortunate enough to sample his creations.

But now, the television's glow only served to illuminate the chasm between his past brilliance and his present ruin. Every perfectly plated dish was a taunting reminder of the man he used to be and the wreck he had become. His fingers twitched involuntarily, a muscle memory of skills now buried beneath the weight of his addiction. These hands, once steady and skilled, now trembled with the echoes of the past, their movements awkward and foreign, like an untrained novice trying to grasp the intricacies of a complex equation.

Jaxon slumped back onto the couch, the oppressive atmosphere pressing down on him like the weight of an atmosphere in a black hole's gravity well. The mid-afternoon light filtered through the grime on the windows, casting long shadows that seemed to stretch infinitely, swallowing him whole. He could feel the distance growing between the man he had been and the wreck he had become–an ever-widening chasm that threatened to engulf him completely. His friends, their concern dismissed, became mere specters, their voices no more than faint echoes in the void.

Inside him, a storm of emotions raged, a chaotic blend of guilt, regret, and the faintest flicker of hope. Guilt gnawed at him like a parasite, eroding the memories of his former
greatness, the celebrations of culinary mastery that had once defined him. He remembered the accolades, the clinking of glasses filled with fine wine, the laughter of friends and admirers. These memories now felt like relics from a parallel universe, glimpses into a reality that had slipped through his fingers like sand.

And yet, beneath the layers of guilt and despair, there remained a single, fragile ember–a mathematical impossibility in a world governed by entropy. It was a reminder that, while distant, change was not entirely out of reach. But in the dim light of his apartment, these thoughts seemed absurdly ambitious, the fantasies of a man who had lost all hope. The ghosts of his friends' voices lingered in his mind, taunting him with their clarity, unattainable and distant.

The afternoon light continued to fade, the encroaching dusk bringing with it a sense of finality, mirroring his internal desolation. Jaxon's mind remained tormented by the contrast between his vibrant past and his hollow present. The dimly lit corner of his apartment, cluttered with relics of his former life and fragments of what could have been, stood as a testament to his internal chaos. He sat motionless, engulfed by the suffocating realization that the man in those photos, the chef who had once commanded kitchens with a fiery passion, was slipping further into oblivion.

Jaxon sat at his makeshift desk, the dim light of the mid-afternoon filtering through a grime-covered window, casting a pallid glow across the cluttered remnants of his former life. His fingers traced the worn edges of an old photo album, the once-vivid cover now dulled and frayed, a relic of a time when life held meaning and purpose. He flipped it open, and the images assaulted him with a cruel
vividness, each one a frozen moment in time, preserved like insects in amber.

The photographs told the story of a man who had once stood at the pinnacle of his craft–a chef whose name had been synonymous with innovation and excellence. In one photo, he was accepting an award, the medal gleaming in the soft stage lights, a testament to his skill and creativity. His eyes lingered on a snapshot of himself smiling triumphantly, holding a golden trophy above his head at a grand culinary event. The applause of adoring fans seemed to echo faintly in his mind, drowned out by the oppressive silence of his current existence.

Each page revealed another scene of success: glamorous restaurant openings where guests dined under extravagant chandeliers, each taste a masterpiece that spoke to his once-boundless creativity. His mind flashed back to the fevered excitement of creating a unique dish for a prestigious food festival. He recalled the precise plating, the intricate balance of flavors and textures that had won him accolades. The memory of pride and joy was now an aching void within him, a reminder of a passion that had been smothered under the weight of addiction.

The apartment around him was a stark contrast to these memories, its walls peeling like the layers of an onion, revealing the decay beneath. The furniture was coated in a layer of dust that seemed to mock the once-vibrant life it had sheltered. Discarded notes and crumpled papers littered the desk, detailing culinary innovations that had never seen the light of day. The lingering scent of stale air, mingled with the acrid stench of cigarettes and alcohol, created an atmosphere of decay that seemed to seep into his very soul.

His fingers hesitated over a photo of a celebratory night with friends and family. In the image, their faces were filled with laughter, glasses raised in joyful toasts. The memory transported him to that night, where the room had been alive with music, the clink of glasses, and the warmth of familial love and camaraderie. It was a stark, painful contrast to his present state of solitude, where the only sounds were the ticking of the clock and the occasional creak of the old apartment.

Internally, Jaxon wrestled with a cacophony of emotions. The pride and fulfillment that had once defined him now felt like cruel jokes played by a callous universe. He grappled with the gut-wrenching realization that the joy he had once found in his culinary art was now an insurmountable chasm away. The very hands that had crafted culinary wonders now trembled with addiction, the haze of substances blurring the brilliance they had once held. He questioned whether he could ever reclaim that part of himself or if he was forever condemned to this shadowy half-existence, trapped in the detritus of what he had once been.

Tears welled up in his eyes, the warmth of the memories burning like acid in the cold reality of his present. The photo album's pages grew blurry as mist gathered in his eyes, reflecting the light of his solitary desk lamp, which flickered intermittently as if sharing in his sorrow. The tears spilled over, each one marking the loss of a connection to his craft and to the people who had celebrated his successes, now distant and seemingly unreachable.

His thoughts drifted to the broader culinary world that had once revered him. The local food scene had been transformed by his daring innovations, his name synonymous with avant-garde excellence. Now, the same society that had once celebrated his heights turned a blind eye to his depths, the stigma of his addiction smothering the brilliance of his former achievements. He felt the sting of societal judgment, the hidden whispers behind sympathetic facades, the quiet condemnation in the silence of their absence.

Fumbling to close the album with a heavy sigh, Jaxon felt the weight of the past settling like lead in his chest. The finality of shutting the book echoed in the room, a symbolic sealing off of a glorious chapter that now felt like a cruel fantasy. The sound of the album closing reverberated through the stillness, marking the end of a journey through time that left him more desolate than before.

The afternoon light continued to fade, the encroaching dusk bringing with it a sense of finality, mirroring his internal desolation. Jaxon's mind remained tormented by the contrast between his vibrant past and his hollow present. The dimly lit corner of his apartment, cluttered with relics of his former life and fragments of what could have been, stood as a testament to his internal chaos. He sat motionless, engulfed by the suffocating realization that the man in those photos, the chef who had once commanded kitchens with a fiery passion, was slipping further into oblivion.

His mind wandered back to the early days of his career, a time when every dish was a new discovery, every flavor combination a revelation. He had lived and breathed his craft, finding beauty in the mathematical precision of a recipe and the chaotic creativity of its execution. The kitchen had been his universe, a place where he could bend the laws of nature to his will, transforming simple ingredients into works of art. But now, that universe had contracted into a singularity of despair, its infinite potential crushed into a point of no return.

As he sat there, lost in the memories of what once was, a profound sense of loss washed over him. It was not just the loss of his career or his reputation, but the loss of something deeper–his connection to the very essence of life. The joy he had once found in creation, in the act of bringing something beautiful into the world, was gone, replaced by an all-consuming void. He realized with a sinking heart that he was no longer the man in those photographs, no longer the chef who had inspired awe and admiration. He was a ghost, haunting the ruins of his own life.

The room around him seemed to close in, the walls pressing against him with a suffocating weight. The air grew thicker, the darkness deeper, as if the apartment itself was absorbing his despair, reflecting it back at him in a never-ending loop. He felt the pull of the past, the weight of his failures dragging him down into the abyss. And yet, somewhere deep inside, a small part of him resisted, clinging to the hope that maybe, just maybe, he could find his way back.

But that hope was fleeting, a flicker of light quickly snuffed out by the overwhelming darkness. Jaxon knew that he was on a path of self-destruction, one that he had chosen willingly, even if subconsciously. He had allowed his addiction to consume him, to strip away everything that had once given his life meaning. And now, he was left with nothing but the hollow echo of his former self, a man who had lost everything, including his own soul.

With a heavy heart, Jaxon closed the photo album and set it aside, the weight of his past settling like a shroud over his shoulders. He knew that he could not go on like this, that something had to change. But he also knew that he was too far gone, too lost in the darkness to find his way back on his own. The thought of seeking help, of reaching out to someone, filled him with dread. He was not ready to face the consequences of his actions, to confront the reality of his situation.

And so, he remained where he was, trapped in the cycle of his own making, unable to move forward, unable to go back. The darkness deepened around him, the room growing colder, the silence more oppressive. Jaxon felt the weight of his solitude pressing down on him, a constant reminder of the choices he had made, the life he had lost. And as the night closed in, he knew that he was alone in this, that no one could save him but himself. But whether he had the strength to do so was a question he was not yet ready to answer.

A knock echoed through the staleness of Jaxon's apartment, a hesitant but persistent sound that disrupted the suffocating silence. He briefly considered ignoring it, but the knock came again, more insistent this time, as though it carried the weight of concern from beyond the void of his isolation. With a groan, he dragged himself from the couch, the leaves of despair rustling with each sluggish movement. He opened the door to find his mother standing there, her face etched with lines of worry, her eyes brimming with the kind of concern that only a mother can carry.

The sinking light from the hallway cast elongated shadows into the cluttered apartment, embedding an eerie stillness that matched Jaxon's hollow interior. His mother stepped in softly, carrying bags of groceries. She smelled faintly of lavender, a scent that seemed almost foreign in the dank, cigarette-stale air that permeated Jaxon's domain. Her eyes

darted around the room, taking in the chaos—the unwashed dishes, the empty bottles, and the scattered paraphernalia. Each item was a symbol of her son's descent, a marker on the downward spiral he had been unable to halt.

"Jaxon," she began, her voice a mix of sternness and deep, aching concern. "I brought you some food."

Jaxon's immediate reaction was one of defensiveness, a reflex born of shame and self-loathing. He clenched his fists, feeling a familiar swell of anger rise within him. "Why are you here?" His voice was sharp, a blade meant to cut through her concern, to push her away before she could see the full extent of his ruin.

She stepped closer, trying to bridge the abyss that addiction had gouged between them, but he recoiled, his anger flaring like a volatile chemical reaction. "You think this helps?" he demanded, his voice rising with a fury that he directed inward as much as outward. His mother flinched, and the groceries almost spilled from her hands. "You think bringing food will fix anything?"

Her face softened, pain interlacing with determination. "Jaxon, I just want to help. Look at you. You're falling apart." She set the groceries down gently, her eyes brimming with unshed tears, each one a silent plea for him to let her in.

Fury surged through Jaxon, a defensive wave of guilt transmuted into aggression. "Help? You can't help me!" He swiped at the bags, sending them tumbling to the floor. Apples and cans rolled under the table, but his mother did not move, rooted by sorrow and a mother's instinct.

"You were everything to us. To me. To your friends. We miss you, Jaxon. Come back to us. Please, seek help." Her words struck at the core of his fragile self-worth, and he turned away, unable to face the pleading in her eyes. Her voice, now a desperate whisper, tried to reach him, but he had already shut the door on any possible connection.

He spat out his response, cold and detached. "I'm beyond help. Just go away."

Tears silently streaked down her face as she gazed at her son, the brilliant chef she once knew now buried beneath layers of despair and denial. She left quietly, her footsteps fading into the hallway, carrying the silence back into the apartment. As the door clicked shut, the weight of solitude bore down relentlessly on Jaxon. The groceries remained scattered across the floor, a testament to his rejection and inner turmoil. His defenses crumbling, he collapsed onto the floor, the darkness of the room seeping into his very soul.

For a moment, he sat in silence, feeling the echo of his harsh words, the crushing weight of his mother's sorrow. The pain behind his eyes intensified, and finally, he gave in. Tears spilled freely as he sobbed quietly, naked in his despair and isolation. With each sob, the reality of his choices and their consequences hit him harder. Memories of laughter and love felt like distant stars, unreachable from the murky depths in which he now resided.

His chest tightened, overwhelmed by the realization of how far he had fallen, the connections he had severed for the fleeting solace of substances. Moments stretched into eternity as he remained on the floor, surrounded by the unopened groceries, now symbols of his lost connections. His tears merged with the dim light of dusk, his cries punctuated by the deep silence of his apartment, resonating through the darkening room where hope felt like an alien concept. The chasm of loneliness was laid bare, making him acutely aware of the void that addiction had carved into his life.

Jaxon Keller sat alone in the dim confines of his decrepit apartment, the stale air heavy with the rancid mix of cigarettes and alcohol. He rolled a joint with mechanical precision, his fingers working through the familiar motions as his gaze fixated blankly on the peeling wallpaper. His mind was a labyrinth of chaos, each twisted corridor echoing with memories of what once was and the hollow silence of what now remained. The spark of the lighter flickered weakly, much like the waning embers of his own spirit.

He took a deep drag, filling his lungs with smoke that swirled and dissipated into the musty air. Jaxon's thoughts meandered through fragments of his past glory, where the sizzle of perfectly seared meat and the aromatic symphony of spices had once painted his world in vibrant hues. Now, those memories felt like ghostly apparitions, tormenting and unattainable, distant echoes of a life he could no longer touch.

An unexpected sound cut through his haze—a distant chorus of laughter. Jaxon stirred, walking towards the window, each step heavy with the weight of disconnection. He peered through the grimy glass at a group of friends gathered around a bonfire in the communal yard. Their joy pierced the fabric of his isolation, reminding him of the fellowship he had forsaken. The sight stirred a tormenting cocktail of envy and regret, twisting his yearning into something almost palpable.

Jaxon's hand trembled as he turned away from the window. His mind warred with itself, a desperate need for human connection conflicting with an overwhelming dread of vulnerability. He stumbled towards the cluttered kitchen counter, his eyes falling upon the relics of his culinary triumphs now reduced to meaningless clutter. Broken trophies, yellowing news clippings, and expired spices gathered dust, mocking the shell of a man he had become.

Driven by a fleeting glimpse of hope, he fumbled for his phone and scrolled aimlessly through his contacts. Names of once-close friends and colleagues flashed by, each tugging at a frayed strand of his conscience. He hesitated over the name of his old sous-chef, a trusted ally in his days of culinary brilliance. His thumb hovered over the call button, his heart racing as doubt and desire collided in a battle that raged within him.

With a deep breath, he pressed the button. The phone rang, each tone resonating with a mixture of anticipation and dread. Suddenly, overwhelmed by the fear of rejection and the shame of his current state, he cursed quietly and hung up before the call connected. The feeling of relief was fleeting, quickly replaced by an even deeper plunge into despair. The silence that followed was suffocating, a stark reminder of his self-imposed exile.

Stumbling back to the couch, Jaxon grabbed a bottle of whiskey and poured a generous measure into a dirty glass. His hands shook as he gulped down the fiery liquid, its warmth a cruel mockery of the comfort he craved. He collapsed onto the couch, tears streaming unchecked down his face. The weight of his past decisions bore down on him, each choice a brick in the wall isolating him from the world outside.

He gripped the phone as if it were a lifeline, clutching desperately to the notion that he was not completely lost. As the alcohol clouded his mind, the pressure of his memories pressed harder, the faces of his loved ones blurring with the faces of those he had failed. His former life, filled with creative passion and human connection,

now felt like a distant planet, unreachable and shrouded in hazy nostalgia.

The laughter from outside seeped through the walls, amplifying his sense of solitude. Society's unspoken expectations loomed large, the chasm widening between who he was supposed to be and who he had become. The voices of his past echoed in his mind, blending into a cacophony of unfulfilled promise and endless yearning.

As sleep began to pull him under, his tear-streaked face relaxed into a semblance of peace. The couch, his prison and sanctuary, cradled him in its worn embrace. Jaxon drifted into an uneasy slumber, the phone still clutched tightly in his hand, a symbol of both his despair and the faint, flickering hope that perhaps, one day, he might find redemption.

Chapter 2

THE UNRAVELING AND THE REBIRTH

Jaxon meanders through the urban maze, a joint precariously held between his fingers, each drag misting his vision with a haze that dulls reality. The twilight sky churns with brooding clouds, casting an uneasy glow over the gritty streets. Flickering neon signs juxtapose the murky air, painting shifting shadows on crumbling walls adorned with vibrant graffiti–the street art of forgotten rebels. The city breathes around him, a living organism pulsating with an electric energy that seems to echo the turbulent state of his mind. The murmur of the city forms a cacophony, a symphony of life teetering on the edge, much like Jaxon himself.
His steps are unguided, his mind a labyrinth of fleeting thoughts, memories of lavish kitchens interspersed with images of bottles and needles. His feet carry him forward, driven by an inertia that feels both aimless and fated. The aroma of street food úmingles with the scent of rain, a prelude to the storm's arrival. Jaxon's once-brilliant mind, now a foggy expanse, can barely register the shift in the

atmosphere as dark clouds mass above, their ominous gathering unnoticed by the passersby who rush to shelter.

The urban landscape is a reflection of his inner turmoil–streets lined with the debris of dreams, alleys leading to dead ends, the occasional flash of neon illuminating the darkness but never quite dispelling it. The city is a maze without a center, a puzzle without a solution, much like the life he now drifts through. The chaotic symphony of honking horns, distant sirens, and the low hum of conversations blends into a white noise that seems to echo the disarray in his soul.

He pauses under a streetlamp that casts a dim, unreliable circle of light, highlighting the cracks in the pavement and the tattered edges of his weathered jeans. He takes another drag, the ember flaring like a desperate heartbeat, and exhales with a hollow sigh, smoke curling like ephemeral whispers. The glow of the joint is a fleeting beacon in the encroaching darkness, a reminder of life's impermanence. Thunder rumbles in the distance, a deep, celestial growl that goes unheeded by the chef-turned-wanderer. His thoughts swirl in a mire of intoxication, rendering him oblivious to the encroaching tempest.

His gaze drifts to the graffiti on the walls, vibrant colors that contrast sharply with the drab surroundings. Each piece of art tells a story, some of rebellion, others of loss, all of them speaking to a life lived on the margins. Jaxon sees himself in those images–once vibrant, full of life and potential, now faded and crumbling, a relic of what once was. The rain begins to fall, soft at first, then harder, each drop a staccato beat on the pavement, merging with the distant thunder to form a chaotic rhythm.

Without warning, a bolt of lightning slams into the nearby street, a blinding lance that sears the sky and earth. The sudden brightness etches shadows against the buildings, capturing a snapshot of the city's chaos. Jaxon startles, his heart pounding erratically, the joint falling from his fingers to sizzle in a puddle. The air crackles with ozone, a scent sharp and metallic, cutting through the haze of smoke that clings to him.

Instinctively, Jaxon tries to seek cover, each movement sluggish, his mind rebelling against clarity. The rain, now a torrential downpour, soaks him to the bone, but it is the second bolt that truly shatters his world. In a heartbeat, it strikes him directly, a heavenly spear thrust into mortal flesh. Electricity courses through his body, every nerve igniting in a violent dance of convulsion. His muscles contract painfully, his vision fracturing into kaleidoscopic shards of light and darkness. He collapses onto the cold pavement, a marionette with severed strings.

As he lies unconscious, the storm roars above, an ominous symphony of thunder and rain. The urban street, once a backdrop to his aimless drifting, now cradles his broken form amidst the tempest's fury. Lightning flickers sporadically, painting the scene in ghostly hues, reflections of a world that teeters on the edge of the real and the surreal. The city, oblivious to the drama playing out on its streets, continues to pulse with life, a machine that devours all who dare to stand still.

Jaxon's body lies still, the rain starting to pelt down in earnest, its cold fingers seeping into his clothes, washing away the grime of his existence. The electric burn leaves a faint, eerie glow on his skin, as though the universe itself has marked him for transformation. The cacophony of the city recedes, blurred by the relentless downpour, as if acknowledging that something profoundly altered lies in the wake of the storm. The concrete beneath him, once a symbol of the urban jungle, now feels like the cold slab of a sacrificial altar, where the old Jaxon is offered up to the forces of fate.

The dimly lit urban street, a testament to both his fall and the precipice of his rebirth, holds its breath. In the throes of the storm, amidst the symphony of chaos, Jaxon Keller's journey into the unknown begins. The storm has passed over him, but it has left its mark–an indelible brand that will guide him on a path he never could have imagined.

Jaxon gradually becomes aware of his surroundings. His eyelids flutter open, and he blinks against the oppressive brightness of the hospital room. The sterile whiteness envelops him, starkly contrasting with the hazy darkness that previously clouded his mind. Everything seems overwhelming, from the glaring overhead lights to the antiseptic smell cloying at his senses. The light, harsh and unyielding, feels like an interrogation, forcing him to confront the stark reality of his existence.

The faint beeping of medical machines echoes in the background, a monotonous rhythm keeping time with his faltering heartbeat. His head pounds, each throb synchronized with the machines. He hears the soft murmur of nurses discussing his condition, their voices a soothing yet incomprehensible hum. Fragments of phrases reach him: "patient," "lightning strike," "remarkable recovery." The words swirl around him, disconnected and surreal, like a language he once knew but has since forgotten.

As he tries to process these sounds, Jaxon attempts to move. A tingling sensation courses through his body, starting at his fingertips and spreading like wildfire. It feels profound, as if each cell in his body is awakening from a deep slumber. This foreign yet electrifying sensation leaves him both exhilarated and terrified. He flexes his fingers, watching them move as though controlled by an outside force, detached yet undeniably his.

His surroundings gradually come into sharper focus. The walls, once plain, now seem to pulsate with energy. Flashing geometric patterns dance before his eyes, intricate fractals spiraling endlessly. He feels a strange connection to these shapes, an intuition that they hold secrets about the universe. His mind races, a torrent of thoughts and images flooding his senses. Each pattern unfurls like a cryptic message, hinting at truths just beyond his grasp.

The beeping of the machines intertwines with the fractal imagery, reinforcing the surreal nature of his new reality. It's as if the patterns and sounds are weaving a complex tapestry, an auditory and visual symphony that speaks to his very soul. Jaxon concentrates, trying to draw meaning from this sensory overload. His heart races, and he feels an insatiable curiosity igniting within him, sparking the first flickers of a journey towards enlightenment.

As the moments stretch, Jaxon's thoughts drift to his past life. He remembers the heat of the kitchen, the rhythmic chopping of vegetables, and the sizzle of ingredients hitting a hot pan. These memories are now intertwined with the dazzling fractals before him, creating a kaleidoscope of his former and present self. Cooking once offered him a semblance of control amidst chaos, but now, his consciousness swirls in an ocean of infinite possibilities.

The tingling sensations intensify, reminding him of the calamity that has led to this metamorphosis. Lightning—a force of nature capable of destruction and rebirth–has transformed him. It feels like a cosmic tuning fork, its vibration realigning his very essence. His heartbeat begins to sync with this newfound rhythm, each beat a step away from the abyss of addiction and a step closer to understanding the interconnectedness of all things.

Jaxon glances toward the door, noticing the silhouettes of nurses moving past. Their hushed tones carry a clinical detachment, a professional distance that makes him feel alien, like an explorer in an uncharted land. "Vitals steady, neurological function..." their words blur into an auditory haze. He wants to speak, to ask questions, but the words elude him. There's an instinctual yearning within him, a desire for connection, yet he finds himself increasingly detached from the mundane world they inhabit.

He becomes acutely aware of his isolation, a stark contrast to the thrumming, interconnected patterns filling his vision. Memories of his family and friends surface, bringing a pang of regret and longing. The fractures in those relationships seem more pronounced now, as if the lightning strike illuminated every fault line in his life. He feels the weight of his past failures, the estrangement from his sister Angelica, and the insidious grip of addiction that once held him captive.

As the overwhelming sensations and images continue to swirl, Jaxon takes a deep breath and allows himself to sink back into the hospital bed. The flood of vivid, spinning shapes and colors is almost too much to bear, yet he finds a sliver of peace within the chaos. He knows this is just the beginning, a profound shift from the depths of despair to the brink of enlightenment. His mind buzzes with the potential to explore these new dimensions, driven by a burgeoning sense of purpose.

For now, he lies still, overwhelmed by the vivid images and sensations, yet quietly determined. The lightning strike wasn't just an accident–it was a catalyst, a cosmic intervention. And as confusion gradually gives way to curiosity, Jaxon feels the first true stirrings of hope. The patterns that fill his vision begin to feel less chaotic and more like a map, guiding him toward something greater, something that lies just beyond the horizon of his understanding.

Jaxon's eyes flutter open, adjusting to the sharp contrast of the brightly lit hospital room against the memories of the storm-laden streets. He tries to rise, but everything seems different–every sensation amplified, every color more intense. The once mundane walls now pulse with intricate patterns, undulating with an energy that feels almost alive. He blinks, then blinks again, trying to dispel what he first assumes are the remnants of his intoxicated mind. But the patterns remain, vibrant mandalas and fractal images swirling on the white canvas of the room.

Each breath he takes feels like he's inhaling the essence of the room itself. The soft chatter of nurses outside his door merges into a harmonious frequency, a symphony of notes that resonates through his very being. He's no longer merely hearing their voices; he's experiencing them as waves that caress his auditory senses and vibrate in harmony with his heartbeat. It's as if reality has peeled back a veil, revealing a world where sound and sight coalesce into something resembling a cosmic dance.

His senses stretch further. A vibrant burst of color floods his peripheral vision, making each hue appear as though it emits its own sound, creating a rainbow of resonating tones. The sterile hospital lights morph into cascading prisms of light, filling the space with spectral rainbows. His fingers, brushing the rough, starched sheets, spark tingling sensations that echo through his nerves, connecting his internal world to the external in ways he's never felt before.

He focuses on a single point at the ceiling where the hospital's sterile white meets the edge of a buzzing fluorescent light. Here, the interplay of shadows and colors forms what looks like an unfolding geometrical pattern, expanding and contracting like a breathing organism. Every tiny movement around him–an unnoticed drip from an IV, the shuffled steps of passing nurses–transforms into elements of this visually acoustic landscape. His mind attempts to make sense of these new dynamics, spinning, churning, forming connections that feel as natural as they are profound.

Startled yet enthralled by these discoveries, Jaxon grapples with a burgeoning sense of curiosity. Each moment brings a new revelation, a deeper layer of his altered state. He remembers his days as a chef, the meticulous care and creativity he poured into every dish, each ingredient speaking a language of its own. Now, he can see those same principles reflected in these surreal patterns, a synchronistic language of existence. Taste, touch, sound, and sight meld together, reborn in this new realm he inhabits.

Still, isolation gnaws at him. The nurses' fleeting presence only accentuates his solitude, as if he is navigating this rewritten world alone. He yearns to communicate, to share these epiphanies, yet words seem inadequate, too blunt for the delicately interwoven fabrics of perception he now experiences. Instead, he sinks back into the bed, letting the sensations wash over him, both comforted and overwhelmed by their intensity.

Driven by the overpowering urge to express what he's uncovering, Jaxon's mind races with possibilities. He wants to map these experiences, to translate this new language of the universe into something tangible. A rekindled fire of exploration ignites within him, reminiscent of the passion he once had for culinary arts. Each dish he created was a work of art, a harmonious blend of flavors and textures. Could this new understanding enhance his creative endeavors, transforming them into conduits of cosmic knowledge?

Jaxon sits up abruptly, every nerve aflame with potential, every synapse firing with a purpose. The bed feels more like a launching pad, poised to thrust him toward this newfound journey. His eyes trace along the fractal patterns one last time, each beat of his heart syncing with the symphony of lights and sounds around him, binding his internal quest with the world's infinite possibilities.

The hospital room fades into the background as his mind locks onto the universe's vastness, ready to embark on this extraordinary exploration. The patterns in his vision grow clearer, more defined, as if guiding him toward a deeper understanding of the interconnectedness of all things. He realizes that this is not just a journey of recovery but one of discovery–of himself, of the universe, and of the infinite possibilities that lie within the spaces between.

Nurses drift into Jaxon's hospital room like ghostly apparitions, their faces framed by the stark, white clinical light. Their soothing murmurs punctuate the sterile silence, but to Jaxon, these sounds are distant, belonging to an alien world. He tries to form words, to explain the overwhelming visions that assault his mind, but his attempts are drowned in their language of routine and monotony. They check his vitals, adjust his IV, and record numbers on a chart, completely unaware of the cosmic upheaval brewing within him. Jaxon watches their methodical actions, feeling like a spectator at an event that scarcely concerns him.

His body throbs with a strange vitality, the electric fingerprints of the lightning strike resonating within his veins. As the nurses exit, satisfied with their assessments, Jaxon's thoughts spiral inward. He recalls the intricate dance of saucepans and chopping knives, the burst of flavors forged in the heat of the kitchen. He visualizes the dishes he crafted, each ingredient meticulously chosen to create a symphony of taste. These memories intertwine with the vivid fractals now consuming his perceptions, each culinary creation morphing into geometric patterns of breathtaking complexity. It dawns on him that his culinary artistry might be the bridge–an expression of the same fractal elegance now filling his consciousness.

He sits up slowly, the sterile sheets rustling beneath him. There is an undeniable connection between the kaleidoscopic visions swirling through his mind and the artistry he once practiced with such fervor. Jaxon's reflections deepen, pondering how the ephemeral nature of taste and texture might align with these cosmic revelations. His thoughts meander like tendrils through the vast expanse of his mind, grasping at the idea that his past not only informs his present but could guide his future. Each dish he ever created becomes a chapter in the fractal manuscript laid bare before him.

Determined to harness these insights, Jaxon contemplates the practices that might help him delve further into this mysterious realm. Meditation beckons, a pathway to exploring the contours of his altered consciousness. He recalls fragmented memories of ancient techniques–breathwork that slows time and expands it, inward journeys that offer glimpses into the heart of the multiverse. He resolves to dive deeper, to embrace the mental pilgrimage that has been thrust upon him.

A surge of empowerment courses through him, igniting a resolve he hasn't felt in years. The lightning strike was no mere accident; it feels like an initiation, a cosmic catalyst propelling him toward a higher purpose. Jaxon vows to break free from the shadows of his addiction, to bury the fog of drugs and alcohol that clouded his once brilliant mind. He clenches his fists, feeling the unfamiliar tingle of renewed determination. No longer will he be a prisoner of his past–a promise forms within the solitude of his thoughts, a commitment to this new path of enlightenment.

Jaxon lies back, the hospital room's fluorescents casting stark shadows on the walls, their cold light failing to dim the fire ignited within him. He gazes at the ceiling, seeing beyond the mundane surface to the fractal tapestry unfolding in his mind. Geometric patterns shift and morph, teasing him with the potential of hidden dimensions just within reach. The nurses' mundane reality fades into the background, their clinical concern overshadowed by the profound transformation brewing within him.

He finds himself both empowered and vulnerable, teetering between the pull of old habits and the promise of new horizons. The nurses' detached care stands in such stark contrast to the cosmic revelations revealing themselves to him. His struggle to articulate even a fraction of this internal cosmos leaves him feeling profoundly isolated, yet their world seems trivial against the immensity of the universe now within his grasp.

A resolute calm settles over him. Jaxon knows that his journey is only beginning. He has glimpsed the infinite, and now, he must navigate this strange new terrain. The paths he once walked and the flavors he once conjured will guide him, but the destination is uncharted. Jaxon closes his eyes, the remnants of the storm still echoing in his ears, and surrenders to the journey ahead.

Chapter 3

BETWEEN FIRE AND FLAVOR

Jaxon sits cross-legged on a mat in the quiet meditation studio, the soft ambient light casting gentle shadows that dance across the room. The atmosphere is steeped in tranquility, a space carved out of time where the outside world seems to fade into irrelevance. His breath slows and deepens, each inhale and exhale becoming a deliberate act of grounding. Around him, the stillness is almost tangible, a cocoon of serenity that silences the chaotic echoes of his past. He shuts his eyes and focuses inward, feeling the rise and fall of his chest synchronize with the rhythm of his thoughts, gradually fading into a tranquil void.

As Jaxon sinks into his meditative state, colors begin to pulse and swirl behind his closed eyelids. At first, they are faint, nebulous wisps, mere hints of what is to come. But as his focus deepens, they explode into vibrant fractal patterns, spiraling and unfolding in an intricate dance. These geometric shapes twist and turn, each one birthing a myriad of smaller, equally complex figures. The patterns are hypnotic, drawing him deeper into their dazzling labyrinth. The vibrant hues and shapes are more than just visual phenomena; they resonate with a cosmic symphony that feels both ancient and profoundly personal.

Jaxon feels an undeniable pull, as if the patterns themselves are guiding him into their core. He is no longer just an observer but an integral part of the unfolding tapestry. The fractals envelop him, and he perceives an overwhelming sense of unity. The designs around him reveal connections between the micro and the macro, the individual and the universal. He recognizes within these fractals reflections of his own mind, each twist and turn a metaphor for his thoughts and experiences.

In this moment, Jaxon grasps a fundamental truth that transcends anything he has ever felt. The universe is not a collection of isolated parts but an interconnected web of existence, each element influencing and reflecting the others. The fractals symbolize the complexity and harmony of this web, mirroring the inner structures of his psyche and the cosmos. As he loses himself in the fractal labyrinth, he feels the weight of his past lift, replaced by a newfound purpose and clarity.

Memories of his former life as an avant-garde chef surface, not as burdens but as essential threads in the fabric of his existence. Each culinary triumph and every dark moment of addiction are critical steps leading to this revelation. The patterns remind him that these experiences, as fractured as they seemed, are all parts of a larger, cohesive design. The shame and guilt that once shrouded his mind dissipate in the face of this realization, making way for a profound sense of acceptance and responsibility.

He recalls Maya, a figure both inspiring and challenging. Their dynamic—a blend of rivalry and mutual quest for enlightenment—comes into sharp focus. Her ambition and dedication to spiritual exploration align with his newfound path, yet he worries that this pursuit could distort their search for truth if tainted by ego. He feels a connection to her that is both profound and inevitable, bound by the lessons and growth they both need to navigate the fractal nature of reality.

Emerged from the meditation, Jaxon's consciousness buzzes with newfound awareness. He understands that the fractal patterns are not just abstract designs but the universe's way of communicating its deepest truths. They reflect the inherent harmony and chaos, light and shadow, that coexist within him and the cosmos. It is a delicate balance, one that demands both respect and curiosity.

Filled with this insight, he opens his eyes, seeing the world anew. The ambient light of the studio now seems to shimmer with a deeper, more meaningful glow. He is part of a grand, intricate creation, a tapestry where every thread, every fractal pattern, contributes to the infinite story of existence.

As he gathers his things to leave the studio, Jaxon feels a subtle shift within himself. The person who entered this space is not the same one who will walk out. His mind, once clouded by confusion and regret, now hums with the potential of what lies ahead. Each step he takes is imbued with the understanding that he is not merely a participant in life but a co-creator, his actions and thoughts rippling out into the vast fractal web of the universe.

The world outside the studio is alive with possibilities, each moment an opportunity to explore the connections he has glimpsed in his meditations. The journey ahead is daunting, filled with uncertainties and challenges, but Jaxon no longer feels weighed down by fear. Instead, he is driven by a quiet determination to uncover the deeper truths of existence, to align his life with the patterns that have revealed themselves to him. The path is clear, if not easy, and Jaxon is ready to walk it.

Jaxon stands before the mirror in his cluttered apartment, the room a chaotic mosaic of forgotten mementos and lingering shadows of his past. The mirror, cracked at the corner from some long-ago accident, reflects a fragmented image back at him. As he gazes at his reflection, he notices subtle shifts in his expression–the weary lines around his eyes seem almost fluid, as if his face is an ever-changing landscape.

In this fractured reflection, Jaxon sees not just his physical form but layers of himself–the vibrant artist he once was interwoven with the confused individual he has become. His past as a celebrated chef, his fall into addiction, and this new, bewildering chapter of his life seem to battle for space within him. The scar on his cheek feels like a dividing line, a marker of a life lived in halves, each side pulling him in different directions.

His eyes, once bright with the passion of culinary creation, now seem to hold galaxies within them, swirling and pulsating with a new understanding. Behind the green orbs, he perceives the ghostly remnant of addiction's grip, juxtaposed against the vibrant patterns that dance within his mind–a kaleidoscope of fractals, each one an echo of the universe's intricate design. The lines of his face, once a simple reflection of time's passage, now seem to map out the journey of his soul.

Deciding to reach out, Jaxon picks up his phone, his fingers hovering over the screen before finally dialing Maya's number. The ringing fills the silence of the room, an almost intrusive sound amidst the quiet chaos. When Maya answers, her voice is warm but tinged with concern, "Jaxon?"

He struggles to find the words, his thoughts a tangled web of sensations and newfound knowledge. His voice trembles, "Maya, I... I can't explain it. Everything's different. Every... thing..."

"Butterflies..." He stumbles over his words, realizing the futility of his attempt. How could he convey the depth of his transformation? "It's like... everything isn't just what it seems."

Maya's concerned voice filters through the line, but it feels distant, almost dissonant against the symphony of thoughts in his head. He sighs, "Never mind, Maya. I think you'd have to see it to understand." Frustration laces his tone as he ends the call. Words fail him, leaving a weight of inadequacy and the gnawing fear of isolation. What if no one could comprehend his newly altered reality?

Turning back to the mirror, Jaxon starts to notice the geometric forms superimposed over mundane details. The lines and contours of his face, the simple curve of his jaw, the arch of his brow, all resonate with a mathematical order that feels as though they have always been there, hidden behind the veil of ordinary sight. Everything is connected, every element a part of an intricately calculated dance.

The clutter of his apartment–once an oppressive reminder of his disordered life–now reveals itself in a different light. The discarded bottles, crumpled sketches, and tarnished culinary tools all pulse with a resonating unity, every object a node in a vast lattice of interconnected reality. He sees spirals in the disarray of clothes, mandalas in the stains on the countertops. The very fabric of his existence is woven with these patterns, each one a reflection of the whole.

Jaxon's mind races with a mix of awe and bewilderment. This new lens through which he perceives the world is both a revelation and a burden. He steps closer to the mirror, the reflective surface now seeming like a portal to a deeper understanding. The pulse he feels resonates through his bones, a steady beat that syncs with the rhythm of the universe. It is as if every atom in his body hums in recognition of this newfound interconnectedness.

He steps back, overwhelmed yet curious, his heart racing with the possibilities that lie ahead. His eyes, still fixed on his reflection, now reflect something beyond mere physicality. In them, the fractal patterns twirl, each curve and shape a signature of the infinite. The dissonance within him begins to harmonize, a flicker of hope ignites amidst the remnants of what was.

Jaxon's breath catches in his throat. The mirror no longer shows just a man recovering from addiction. It shows a conduit, a bridge between the known and the infinite. He places a hand on his chest, feeling the steady pulse that connects him to everything around him. The world has not just changed; it has unveiled its true essence to him. He is not merely observing the world; he is participating in its creation, his every thought and action rippling out into the fractal web of existence.

Still facing the mirror, Jaxon steps back, letting the awe and apprehension wash over him. He knows this is only the beginning of a profound journey, one that blends wonder with trepidation. His reflection stares back with eyes that shine not just with knowledge but with the power and responsibility that newfound clarity demands. This journey is not one he can undertake lightly, but it is one he feels ready to embrace.

Sunlight drapes the park in a warm, golden hue, making the leaves tremble in its gentle embrace as though whispering secrets of the universe. Jaxon strolls down the cobblestone path, the air redolent with the scent of blossoming flowers and freshly cut grass. A light breeze rustles through the trees, entwining with the melodic chirping of birds, creating a sensory tapestry that shifts Jaxon between reality and a world shaped by his newly enhanced perceptions.

He pauses beside a bed of flowers. Each petal reveals itself to him as a fractal universe, spiraling infinitely inward and outward, hues of violet and amber blending in a mesmerizing dance of light and shadow. He traces the lines with his eyes, following the patterns that speak to him in a silent, mathematical language. The geometry of these natural forms strikes him deeply–each bloom, an echo of cosmic arrangements, each curve, a whisper of the universe's fundamental design work.

As he moves to a bench, his fingers brush its wooden surface, feeling the grain that no longer registers as mere texture. Instead, he sees intricate, interconnecting pathways that bridge the synthetic with the organic, a testament to the underlying order governing all forms. He sits down, letting the patterns unravel in his mind, where benches transform into interlaced algorithms.

Pulling out his phone, he begins to capture these discoveries. A tree's branches splitting into smaller and smaller forks, serpentine lines tracing the path to the sky–a perfect, biological fractal. The surface of a pond, dappled with ripples, each wave concentric and interconnected, reflecting a higher dimension of quantum entanglements. Jaxon snaps photo after photo, hoping to trap these fleeting moments of enlightenment within the confines of a digital lens.

The photography ignites a deeper need–a need to codify these discoveries, to etch them into permanence. He takes out his sketchbook, hands trembling in reverence. Under his pencil, lines spring to life, coiling and spiraling, morphing into fractal designs that embody both chaos and perfection. His sketches are an ode to unity. Within these shapes, he sees how the structural rules of the universe extend to all of its manifestations, from the simple lines of a leaf's vein to the complex tessellations of galaxies.

Jaxon's thoughts drift to his past, to the bustling kitchens where he once created culinary masterpieces. Cooking, much like these fractal patterns, was a dance of precision and intuition–a balance of order and creativity. Each dish was its own intricate universe, composed of flavors and textures, mirroring the complex and beautiful structures he now perceives in everything around him.

The world of art and mathematics have long been considered separate realms, but in the swirling designs he inscribes in his sketchbook, Jaxon perceives the folly of such distinctions. Art is math, math is art; creativity and logic unify in this new vision that transcends traditional boundaries. In geometric precision, he finds beauty; in the unpredictable curves, he finds artistry. This realization is a cornerstone of his transformed consciousness, challenging societal perceptions and paving the way for a more integrated understanding of existence.

Lost in these reflections, he feels the pull of two worlds—the chef he used to be and the cosmic explorer he has become. Each stroke of the pencil brings him closer to bridging this divide. Can these patterns redeem him, heal the fractures of his tormented past? Or will he remain ensnared by the shadows of his former self? These questions gnaw at him, propelling his hand in a frenetic dance across the paper.

Passersby glance at him curiously, as he mutters to himself, barely perceiving the outside world. They are but ghostly apparitions in his quest, mere background noise to his interior symphony. A child stops to watch, eyes wide with curiosity. Unlike adults, children see beyond the ordinary—a trait Jaxon has been reacquainted with post-awakening. He hands the child a finished sketch of flowers spiraling into infinity and receives a beaming smile in return, a simple acknowledgment of beauty shared and understood.

Jaxon sets down his sketchbook, his mind a tempest of thought and wonder. The patterns he's drawn are incomplete chapters in a larger, grander narrative—chapters he feels compelled to explore and expand. He leans back on the bench, eyes closed, the fractal geometry of existence swirling behind his eyelids. Nature's rhythms pulse in unison with his own heartbeat, each moment a testament to his interconnectivity with all things.

With every revelation and sketch, a sense of purpose begins to crystallize within him. The chef who once delighted in creating sensory masterpieces is evolving. His art is no longer confined to taste and smell; it now encompasses sight and mind, weaving an intricate, cognitive tapestry. This park, with its myriad of natural symmetries, becomes his open-air laboratory, where he rekindles his passion with a newfound depth.

As Jaxon opens his eyes, the world comes alive anew, brimming with the fractal majesty he is now attuned to. He gathers his belongings, feeling an urge to continue his exploration, this time through the medium that once was his sanctuary—the kitchen. The bench stands empty now, save for the lingering echo of his insight, promising a journey of creations bound not just by earthly flavors but by the infinite patterns of the cosmos.

Jaxon sits at his scarred wooden kitchen table, surrounded by a chaos of notebooks, each one brimming with scrawled diagrams and phrases that seem to spill over the edges. His pen hovers above the page, trembling with the unresolved tension of his thoughts. The soft light from a nearby lamp casts intricate shadows, accentuating the deep lines on his face as he stares intently at his most recent drawing. The fractal patterns twist and curl on the paper, alive with a complexity that his mind struggles to fully grasp. He reaches up and rubs his temple, feeling the familiar throb of frustration as words fail him once again.

In his frustration, Jaxon feels the weight of his past pressing down upon him. Once, his genius was celebrated; his culinary creations were a testament to his brilliance. But that same brilliance had been dulled by addiction, leaving him to wade through days blurred by a haze of substances. Now, with newfound clarity, the inadequacy of the language to encapsulate his experiences gnaws at him. Each word seems inadequate, like trying to capture a storm in a glass bottle. He aches to articulate the profound revelations he's encountered, but each attempt feels like a pale imitation of the truth.

Feeling the surge of inadequacy washing over him, his thoughts wander to Angelica. He can trace the lines of their history as if etched into his very being. Her laughter, his protector in childhood, was replaced with tears of disappointment when his life veered off course. Memories of her attempts to pull him from the abyss of addiction haunt him. Her voice had been a lifeline, a persistent reminder of familial love, but he always found ways to let her down. The gnawing guilt of dragging Angelica into his chaotic world compounds his current struggles, adding another layer to his frustration.

Stepping back from his notebooks, Jaxon's gaze settles on the mess of his kitchen. The remnants of another life are strewn around–aprons stained with the colors of culinary experimentation, knives dulled and forgotten. This space, though chaotic, holds a sanctuary-like reverence for him. Here, surrounded by the tools that once made him a notable figure in the culinary world, he feels a flicker of hope. Perhaps if words fail, his cuisine can bridge the gap.

Shaking off the encroaching despair, he shifts his focus. He picks up a sketchbook and begins outlining a new culinary concept inspired by the fractal patterns. The geometric forms he draws seem to sing with life, echoing the structures that have been revealed to him in his meditative states. Every fleck of ink holds a piece of the cosmos, a glimpse into the interconnectedness he's felt so viscerally. His drawings, though abstract, start to take the shape of dishes infused with the essence of his newfound understanding.

Jaxon's hand moves purposefully, capturing the interplay of shapes and flavors. He envisions constructing a dish that is more than a mere arrangement of ingredients. These plates will be metaphors, mirroring the symmetry and intricate beauty of nature's geometry. His movements become almost dance-like, each step imbued with intention and reverence. With every stroke, he resolves to trust his feelings, letting his instincts guide him through the creative process.

The kitchen is filled with the rustle of paper, the scratch of a pen, and Jaxon's rhythmic breathing as he transitions from drawing to action. He moves to the counter, gathering ingredients with newfound determination. His fingers trace the contours of fresh vegetables, feeling their textures and recognizing the patterns in their organic shapes. He begins to slice them carefully, each cut deliberate, his mind connecting the actions to the greater fractal symphony he's come to understand. The scent of herbs fills the air, an aromatic prelude to the culinary masterpiece taking shape.

As he works, his thoughts meander through his past failures and the skepticism he still holds about his ability to transcend them. Yet, amid the slicing and stirring, he feels a shift–a sense of unity forming between his past self and the potential he now embodies. Every action in the kitchen becomes a whisper of redemption, a reclamation of his identity as both a chef and a conduit of higher understanding. The culinary world, once merely a stage for his temporary brilliance, now appears as a sacred vessel for healing and connection.

The kitchen, with its assortment of utensils and the hum of the refrigerator, becomes a canvas for Jaxon's renewed artistic fervor. He meticulously arranges each component of his dish, mindful of how each geometric arrangement mirrors the fractal insights he's had. The final results are not just food but visual testimonies to his mental and spiritual journey.

By the time he finishes, the dish before him is a marvel of visual symphony—a delicate balance of colors, shapes, and textures that not only appease the palate but also narrate his story of transcendence. As Jaxon stands back to admire his creation, he feels an overwhelming sense of purpose. Despite the earlier frustrations and the language barriers that hampered his expression, he realizes that through his cooking, he can communicate the awe and wonder of his revelations. This shift in his creative expression leaves Jaxon both exhausted and exhilarated, resolving to continue this path, using his culinary art as a bridge to convey the cosmic truths he has touched.

Each movement is intentional and mindful, his hands executing the vision his heart has ignited. Jaxon feels the resonance of the universe in every slice, every seasoning, every artful placement of food on the dish. Through his creations, he will offer a taste of the infinite, inviting others to embark on their own journeys of inner exploration. As he prepares the dish with deliberate care and newfound insight, he feels a flicker of hope rekindling within him—a hope for redemption, connection, and the possibility of sharing his deepened understanding of the fractal nature of reality.

Chapter 4

WHISPERS OF THE INFINITE

Jaxon stands in the center of his dimly lit apartment, a sanctuary that lies somewhere between chaos and contrived order. Shadows cling to the ceiling corners like memories of his past, both taunting and whispering promises of renewal. The air is dense with a mixture of remnants–an empty whiskey bottle left on the kitchen counter, a burnt-out candle atop a stack of culinary magazines, and a small, disheveled plant struggling for life in the window. A stark reminder of his former self, the infinity symbol entwined with a fork and knife tattooed on his forearm catches the dim glow of the single desk lamp, a relic of both his culinary genius and his descent into addiction.

The room is more than a physical space; it is a reflection of Jaxon's internal landscape. Every item, every misplaced object tells a story of his journey–his rise, his fall, and his hesitant steps toward redemption. The clutter is not just physical; it represents the tangled thoughts, unresolved traumas, and unspoken regrets that have shaped his life. The walls, marked by time and neglect, seem to close in on him, reminding him of the isolation he has created for himself. This apartment, once a haven of creativity, now feels like a tomb of memories, a place where time stands still.

As he moves through the room, his fingers brush over objects that once held meaning: a chef's knife dulled by disuse, a notebook filled with half-formed ideas, and a photograph of him and Angelica, taken long before addiction had driven a wedge between them. Each touch stirs a memory, some sweet, others bitter. The weight of these memories presses down on him, making the air feel thicker, harder to breathe.

He begins gathering the necessary tools for his DMT journey–an old pipe, a small vial of the crystalline substance, and a lighter. His movements are deliberate, almost ritualistic, as he sets up a guided meditation recording on his laptop. The calm, instructive voice promises to anchor him, a steadying force against the unknown turbulence he is about to willingly dive into. The scent of sandalwood incense begins to fill the room as he lights a stick, the curling smoke weaving through the slices of light that pierce the darkness from the streetlamps outside.

Jaxon's mind flits back to the hospital room where he first awakened after the lightning strike, the sterile scent of antiseptic mingling with the faint aroma of hope. The clarity that washed over him then felt like a cosmic realignment, a shift that nudged his consciousness into a realm where ordinary perception failed. He had emerged from that experience altered, his mind reshaped into a prism refracting the universe's hidden dimensions. Now, as he hovers on the brink of another altered state, hesitancy and anticipation intermingle.

"Focus on your intention," he murmurs to himself, voice barely above a whisper, as he dims the lights further, creating a cocoon of serenity. The soft illumination leaves only what he needs in view–a clean slate for the vivid tapestries soon to unfold in his mind. He takes a moment to breathe deeply, to center himself, knowing that the journey ahead will challenge him in ways he cannot yet comprehend.

As the DMT vapor invades his lungs, Jaxon shuts his eyes, plunging into a kaleidoscope of overwhelming colors and fractals that instantaneously animate his vision. It starts with a burst of electric blue, morphing into spiraling patterns of verdant green and pulsating crimson. The apartment around him dissolves; walls and furniture become inconsequential artifacts in the expanse of his altered consciousness. Breathing deeply, he surrenders to the vivid assault on his senses, seeking the guidance these emerging visions might offer.

His internal monologue is a constant whisper. "What do you need to show me?" The question echoes into the fractals. Flashes of his estranged sister, Angelica, surface sporadically amidst the geometric symphony, her image a delicate fractal pattern in itself. They hadn't spoken much since his addiction had driven a wedge between them, but here, under the influence of this cosmic introspection, memories of their shared childhood resurface–whispers of laughter and warmth, shadows of guilt and a longing for reconciliation.

These memories are not just images; they are sensory experiences. He can feel the warmth of Angelica's hand as they ran through the fields near their childhood home, the sound of her laughter echoing in his ears. He can smell the faint scent of her perfume, a scent that always made him feel safe. But these memories are tinged with the bitterness of regret, a reminder of how far he has strayed from the person he once was. The guilt is palpable, a weight that presses down on his chest, making it hard to breathe.

As the DMT peaks, Jaxon's visions take on corporeal form, swirling entities communicating in a language of symbols and light. Every thought feels as though it taps into the universal lattice of existence, his consciousness folding in upon itself. The message seems to clarify: his life's purpose isn't just in deciphering this fractal reality, but in making amends and reconnecting. He sees his past decisions not as random acts but as interconnected threads, each one a part of a greater cosmic fabric. The weight of his expectations morphs into a realization–illumination must bridge both the internal and the external.

Gradual acceptance filters through him. Feelings of profound peace and clarity replace the initial overwhelm. Symbols evolve into coherent thoughts, and he perceives intricate maps of knowledge, each one offering a new layer of understanding. His vision narrows and focuses on a singular holographic equation, shimmering with an ethereal glow. It is both familiar and foreign, a cosmic blueprint whispering truths about the universe's multifaceted reality.

Jaxon tries to commit its fractal intricacies to memory, even as the visions begin to recede. With each passing moment, he finds himself anchored more firmly back to the present, but the specters of insight into the equation persist, promising more revelations in quieter moments of contemplation. He feels as though he is on the brink of a profound discovery, something that could change the course of his life if only he could grasp it fully.

Finally, he emerges from the psychedelic intensity and sits quietly in the dusky apartment, feeling the echoes of cosmic consciousness rounding back into the confined reality of his room. The warmth of his body grounds him, the vapor dissipating into the air, leaving behind nothing but a faint chemical scent and the lingering taste of the universe's infinite possibilities. His heart beats steadily, a rhythmic reminder of his place within this realm, even as he begins to integrate the vastness of what he has seen.

With the guided meditation still playing softly in the background, Jaxon remains seated, absorbing the aftermath of the cosmic encounter. There is a sense of alignment within him, a quiet determination forming amongst the surreal clarity. Purpose gleams at the edge of his consciousness, urging him onward into the labyrinth of understanding that lies ahead. His fingers trace the contours of the infinity symbol tattoo as if reaffirming his resolve. The fractal images persist at the periphery of his mind, as he sits quietly, his thoughts illuminated by the hues of revelations yet to be fully grasped.

He knows that this is just the beginning. The journey he has embarked upon will not be easy, but it is necessary. He must confront the demons of his past, heal the wounds he has inflicted on himself and others, and embrace the profound truths that have been revealed to him. The road ahead is uncertain, but for the first time in a long while, he feels a sense of hope, a belief that he can find his way back to the light.

Jaxon steps through the doorway of the meditation center, feeling a delicate shift from the urban chaos clinging to him to the sanctuary's calming aura. The air is tinged with the subtle fragrance of sandalwood, guiding him deeper into the space. Soft, natural light filters through floor-to-ceiling windows, casting a golden hue over the grounds covered in lush greenery. The diverse group of participants already immersed in their own meditative journeys greet him with serene glances and gentle nods, creating an atmosphere of welcome and acceptance.

He selects a cushion near the back of the room, hesitating momentarily before settling down. The sounds of the city gradually fade, replaced by the gentle trickle of a nearby fountain and the faintest whispers of instrumental music. His breath slows, matching the rhythm of the room. The instructor's voice weaves through the air, guiding everyone into a deeper state of relaxation.

Jaxon's initial discomfort lingers, a remnant of his recent struggles. Memories of a frenetic kitchen, the clang of metal against metal, and the overwhelming sensory overload tug at the edges of his mind. He must fight to stay present, to let the gentle guidance of the instructor pull him into the realm of stillness. The soothing cadence of the voice encourages him to focus on his breath, to sense the rise and fall of his chest as if it were the ebb and flow of a vast ocean.

As he settles into the practice, his mind begins to quiet. The external noise, both literal and metaphorical, fades into the background, replaced by a deep, resonant silence. This silence is not empty; it is full of potential, a space where thoughts and emotions can surface without judgment. Jaxon feels himself sinking into this silence, allowing it to envelop him, to hold him in a place of safety and calm.

Settling into the rhythm of the practice, Jaxon closes his eyes, sinking into a meditative state. The fragrance of incense mingles with the subtle scent of nature wafting from the garden. As he sinks deeper, the fractal patterns from his DMT trip begin to resurface, dancing behind his closed eyelids in vivid colors and shifting shapes. The patterns unfold like a cosmic kaleidoscope, each twist and turn a reminder of the universe's infinite complexity.

His internal monologue drifts, blending wonder and skepticism. He recalls flashes of his past–his culinary genius, now tainted by addiction. The kitchen mishap that left a scar on his cheek stings anew, a physical marker of his fall. But amid the chaotic memories, the fractal visions offer solace, presenting a glimpse into a reality where time and space are boundless. He yearns to harness this newfound understanding, to reclaim the brilliance he once wielded with a chef's knife but now seeks through the intricacies of his mind.

The guided meditation continues, and he feels a shift within. The swirling patterns give way to a profound sense of oneness. Each breath connects him to an unseen web, where he is but a single node in the grand tapestry of existence. Unity with the cosmos envelops him, dissolving the remnants of his disconnection. For the first time, he feels at home in his own skin, embraced by the universe's fluid embrace.

The session concludes, and the participants slowly open their eyes, the room filling with a collective sigh of release. Conversations sprout like delicate shoots, tentative and gentle. Jaxon finds himself approached by others, their faces reflecting a shared purpose. Together, they discuss their insights, and for Jaxon, it feels like uncovering buried treasure. The others' experiences resonate deeply, each story a mirror reflecting different aspects of his journey.

Jaxon speaks of his DMT trips with a sense of awe, describing the fractal beauty he witnessed. The group listens, their expressions shifting between curiosity and understanding. Maya appears at his side, her presence both comforting and challenging. They exchange glances, filled with unspoken rivalry and camaraderie.

"It's remarkable how our minds can stretch beyond what we perceive as reality," Maya muses softly, her voice carrying the weight of countless philosophical debates.

Jaxon nods, his thoughts aligning with hers yet pushing beyond. "It's as if each experience peels back a layer, revealing a truth that's been there all along, waiting for us."

Their exchange weaves through the tapestry of their shared quest for enlightenment. The group's warmth and encouragement fill Jaxon with renewed vigor. He senses the bond of communal support tightening around him, a lifeline in the vast sea of his exploration.

As he leaves the meditation center, the evening air feels different, charged with a new purpose. He walks with a lighter step, the weight of his past balanced by the profound connection he has cultivated. The fractal patterns linger in his mind, guiding him forward on a path lit by the insights of shared exploration. For the first time in a long time, Jaxon feels a deeper connection not only with himself but with the world around him.

This sense of connection extends beyond the group. As he walks through the city streets, he notices the interconnectedness of everything around him. The way the shadows of buildings dance with the light of the setting sun, the rhythm of footsteps on the pavement, the hum of distant traffic–all of it feels like part of a greater whole. Jaxon feels himself moving in harmony with this rhythm, a small but integral part of the world's intricate dance.

As the night deepens, Jaxon reflects on the journey he has undertaken so far. The struggles, the pain, the moments of clarity–all of them have led him to this point. And while the path ahead remains uncertain, he feels ready to face it with an open heart and a clear mind. The insights he has gained, both from his internal exploration and from the collective consciousness of the group, have given him the strength to continue. He knows that he is not alone in this journey, that he is supported by the wisdom of the universe and the connection he shares with others.

Jaxon sits on a weathered wooden bench in the heart of the tranquil park, the world around him a soft ballet of fluttering leaves and whispering breezes. With his notebook spread open on his lap, he sketches delicate, intricate fractal designs. The air here is different from the disarrayed, suffocating atmosphere of his old haunts. His mind starts drifting, contemplating the essence of time, pondering the intricate structure that binds the universe together.

As his pencil glides over the paper, Jaxon's thoughts twist through the implications of non-linear time–a concept he has often found elusive yet suddenly begins to make sense. Traditional views of chronological progression fade away, replaced by a more fluid understanding inspired by his recent mystical experiences. He recalls theories from modern physics–relativity and quantum mechanics–emphasizing time as an intertwined tapestry rather than a mere sequence of events.

These ideas intertwine with the rich tapestry of indigenous philosophies he encountered through his studies, cultures that perceive time as cyclical, a continuous dance without a definitive beginning or end. He considers how societies governed by linear time cling to schedules and deadlines, missing the boundless beauty in moments' interconnection.

Lost in this whirlpool of thoughts, Jaxon observes the park's life—groups of children chasing each other, their laughter an ephemeral note in the symphony of existence. An elderly couple ambles slowly along a gravel path, their hands entwined, their pace unhurried by the constraints of ticking clocks. Jaxon imagines their lives as complex patterns of past and future, a web of memories and anticipations intricately interwoven with the present.

Each individual in the park becomes a focal point, an anchor in the vast river of time. What lives have they lived? What paths will they travel? Jaxon envisions each life as nodes in a grand, multidimensional lattice, each moment flowing in varied directions, overlapping and merging. His mind, electrified by this idea, begins to visualize these lifelines as rivers intertwining, flowing freely rather than adhering to rigid boundaries.

He contemplates the notion that every choice, every action, creates ripples in this river of time. These ripples interact, merge, and sometimes clash, creating the intricate patterns that define our lives. He wonders how his own ripples have affected those around him—how his actions, both good and bad, have shaped the lives of others. The thought is humbling, a reminder of the interconnectedness of all things.

The tranquility of his surroundings sharply contrasts with the internal chaos Jaxon has battled. The quiet ripple of the pond, the rustle of leaves—all these meld into a serene melody that calms the turbulence within him. He begins to reconcile the disjointed fragments of his past—the trials of addiction, the pain of lost connections, the ethereal whispers of cosmic truth heard under the influence of psychedelics. Each memory, previously a jagged shard, now fits into a more harmonious mosaic, a pattern defined by both chaos and order.

Jaxon's contemplation deepens as he sketches more fractal designs, inspired by nature's inherent mathematics. Each spiral and loop represents infinity within a finite space, a visual metaphor for the interconnected universes within each mind. He outlines mathematical principles in his notebook, finding a strange beauty in the harmony that emerges from chaos. These principles, once abstract ideas, now take concrete form, reflecting his new understanding of time's fractal nature.

In a profound moment of clarity, Jaxon perceives time not as a line but as a river–ever-changing, endlessly flowing. It branches into numerous streams, merging with others, evaporating into the air only to fall again as rain, returning to the ever-present flow. This insight is both humbling and empowering. He feels as if he is standing at the confluence of all times, past, present, and future, sensing the unity within the multiplicity.

Jailed in his worst days, Jaxon was a fragment, estranged from the continuity of life. Now, each breath, each heartbeat, connects him to an eternal rhythm. His fractal sketches, once mere doodles, now manifest as a map–an equation–to navigate the multiple dimensions of being. The realization feels like sun breaking through the clouds, illuminating the intricate patterns of existence he had only glimpsed in his altered states.

Closing his notebook with a sense of fulfillment, Jaxon reflects on his journey. The park around him–a sanctuary of simple joys and quiet reflections–becomes a symbol of his evolving consciousness. Here, in this serene cocoon, he finds purpose, a guiding light in his quest for understanding. The clarity and insight gained today are not fleeting epiphanies but transformative keystones that will pave his path forward.

The simplicity of the park mirrors the profound simplicity of his discovery. The interconnectedness of lives, the river of time, the fractal beauty within chaos–each of these elements align, bringing harmony to his once tumultuous mind. Closing his eyes, he breathes deeply, feeling an intimate connection to the universe and its infinite mysteries.

Jaxon leaves the park with not just a deeper understanding of time but a renewed sense of clarity and purpose. The path ahead may be fraught with challenges, but he feels prepared to face them. The insights he has gained, both from his internal exploration and from the world around him, have given him a new perspective on life. He knows that the journey is far from over, but he is ready to embrace whatever comes next.

As he walks away from the park, he carries with him the lessons of the day–the understanding that time is fluid, that life is interconnected, and that there is beauty in both order and chaos. These lessons will guide him as he continues to explore the mysteries of existence, as he seeks to find his place in the grand tapestry of the universe.

The dense foliage of the Amazon rainforest genuinely envelops Jaxon as he steps into the retreat, the crisp, humid air filling his lungs with each breath. The shaman guides, their presence commanding yet serene, stand at the entrance. They wear simple robes adorned with tribal patterns, their faces etched with wisdom and experience. Jaxon catches their eyes, feeling an immediate connection, a silent recognition of the journey he is about to undertake. He smiles awkwardly, joining the circle of participants, each one here to seek transformation just as he is.

His mind races with a cacophony of thoughts–memories of his time drenched in addiction, suffocating in the shadows of his past mistakes. He tries to banish these visions, focusing instead on his surroundings. The retreat is a sanctuary in the heart of the wild, a secluded hideaway where ancient rituals meet modern seekers. The sound of a distant waterfall mingles with the rustling leaves, creating a natural symphony that calms his racing heart.

The shaman beckons him forward to participate in the cleansing ritual. Sage smoke curls in the air, its pungent aroma sharp and cleansing. Jaxon steps barefoot on the cool, damp earth, feeling grounded for the first time in what seems like forever. The shaman chants softly, the words mysterious yet soothing, bathing Jaxon in an aura of expectancy. He stands still as the shaman waves a feathered wand around him, the smoke enveloping his body like a protective cloak. He sets a clear intention in his mind–a desire to strip away his past and reforge his spirit anew.

As the ceremony progresses, Jaxon senses the old traumas rising to the surface. Visions of nights spent in drunken stupor and days lost in a haze of drugs flood his mind, forcing him to confront the darkness head-on. He feels his stomach knot, the weight of his past errors bearing down on him. Yet, there is also a flicker of resilience, a determination to transcend these shadows and move toward the light.

The shaman's chanting grows louder, more intense, as if urging Jaxon to release the burdens he has carried for so long. The sound vibrates through him, resonating with the deepest parts of his being. He feels the darkness within him rising, swirling, as if preparing to be purged. The intensity of the moment is overwhelming, but Jaxon knows that this is a necessary part of the healing process. He must face the darkness head-on if he is to move forward.

With the ritual complete, Jaxon kneels before a fire pit, the flames casting dancing shadows on his face. The shaman hands him a small wooden cup filled with a thick, muddy brew. The Ayahuasca–his gateway to the unknown. He takes the cup with a trembling hand, the weight of the moment pressing heavily on him. He whispers his intention softly into the liquid before bringing it to his lips, the bitterness shocking his taste buds as he gulps it down. He crawls over to join the others around the fire, feeling the warmth suffuse his body, as if the flames themselves are seeping into his very soul.

Minutes stretch into what feels like eternity. The psychedelic effects begin to unfold like an ancient tapestry, threads of vibrant hues and swirling fractals weaving together to form an intricate pattern in his mind's eye. Jaxon lays back, gazing up at the night sky through the canopy of trees. Stars blur into trails of light, spiraling into geometric shapes that dance across his vision. He closes his eyes, feeling the vibrations of the universe coursing through him, a symphony of existence he can neither fully comprehend nor resist.

The brew plunges him into a vivid, intense state. The boundaries of his physical form dissolve, merging with the cosmic symbols that proliferate in this altered state. He sees memories of his youth, reliving moments of unbridled joy and deep despair, all converging into a single point of awareness. Old traumas take form–specters from his past confront him, demanding acceptance and release. He trembles, tears mingling with sweat on his face, but he surrenders to the experience, allowing the Ayahuasca to guide him deeper into the labyrinth of his psyche.

In this profound state, the messages begin to unfold. Rainbow tendrils of light extend from his body, connecting him to the immense web of existence. He witnesses the interconnectedness of all things–every being, every event, intertwined in a grand tapestry of life. The sensation is overwhelming, a wave of unity that crashes over him, washing away the remnants of his former self. Insights permeate his consciousness—truths about love, forgiveness, and the boundless nature of the soul. He feels his spirit lightening, the heavy chains of guilt and regret dissolving in the cosmic embrace.

As the visions start to ebb, Jaxon finds himself sitting quietly amongst the other participants. The ceremony concludes with a sacred chant, the voices harmonizing in a mesmerizing melody that resonates deeply within him. He opens his eyes to see the others around the fire, their faces illuminated by the flickering flames, each one bearing an expression of awe and comprehension.

He has crossed a threshold, a profound sense of unity binding him to these fellow seekers and to the source of all creation. Gazing into the night, with the jungle whispering its ancient secrets around him, Jaxon realizes the significance of this moment. He has touched a piece of the infinite, and in doing so, has begun to understand his place within it.

The night sky above holds no answers, just an infinite canvas of possibilities, each star a point of light guiding him forward. He feels a newfound clarity, a purpose crystallizing in his mind. No longer is he a mere wanderer lost in the sands of time—he is Jaxon Keller, a soul on the path to redemption, forging a connection between the profound wisdom of the cosmos and the deep mysteries within.

Exhaling deeply, he closes his eyes once more, allowing the experiences of the night to settle within him. The path ahead is unclear, but for the first time in a long while, he feels a sense of peace and possibility. The seeds of enlightenment have been planted, and as he sits among the others, he knows that his journey has truly begun.

Chapter 5

THE ALCHEMIST'S TABLE

In the dim light of his cluttered studio, Jaxon sifts through a sea of notes, each piece of paper a fragment of his psychedelic journey. Symbolic drawings and scrambled equations cover the space, chaotic yet holding a promise of clarity. The air is thick with the scent of incense, a soothing counterpoint to the mental turbulence that engulfs him. His eyes, bloodshot from hours of toil and introspection, scan the myriad sketches, seeking coherence within the chaos.

The studio, once a culinary sanctuary where Jaxon had conceived his most audacious dishes, now stands transformed into a crucible for his fractured mind. The kitchen equipment that once gleamed with purpose now gathers dust, overshadowed by the equations and geometric patterns that seem to pulse with life on every available surface. The scent of incense mingles with the faint aroma of burnt coffee, a lingering reminder of his sleepless nights.

With meticulous care, Jaxon spreads the pages across the floor, creating a mosaic of his fragmented mind. Each sketch speaks to him—visions of geometric patterns seen under the influence of psychedelics, echoes of sensations felt in meditation. His heart pounds with the fervent energy of discovery. This is more than a mathematical quest; it reflects his struggle with addiction and longing for redemption. Every stroke on paper is a step away from the abyss and a stride toward enlightenment.

As he arranges the sketches, his thoughts drift to the moment when this journey began: the lightning strike that jolted him into an altered state of consciousness, opening his mind to the fractal nature of reality. That moment, both terrifying and exhilarating, marked the beginning of his transformation. It was as if the universe had reached out to him, offering a glimpse of its intricate design, a design he now sought to decode.

His confidant, a calm and inquisitive presence, sits nearby, listening intently as Jaxon speaks. "These equations aren't just numbers," Jaxon murmurs, his voice thick with urgency. "They are... mappings. Pathways to something beyond our perception."

"What exactly do you see when you look at them?" his confidant asks, leaning forward, probing gently yet persistently.

Jaxon pauses, his fingers tracing the lines on a sketch of a complex geometric shape, memories of his culinary days flickering through his mind–days when flavors and textures were his canvas. "I see dimensions, hidden layers of reality. It's like... every shape reveals a truth. The universe, it's all connected in this intricate dance, and we've only been glimpsing the edges."

His confidant nods, encouraging him. "But why you, Jaxon? Why now?"

Jaxon's eyes darken with the weight of his past. "Maybe it's cosmic irony," he says, a wry smile tugging at his lips. "From a chef who couldn't handle life's simplest ingredients to... this. I think the lightning strike wasn't an accident. It's as if the universe decided to give me one last chance to find meaning–beyond addiction, beyond despair."

The whiteboard looms in the corner of the studio, pristine and ready. Jaxon stands, chalk in hand, and begins to draw, his movements fluid yet deliberate. Lines intersect and curves unfurl, inspired by the sensations that rippled through his altered states. The room hums with the energy of creation, every mark on the whiteboard a beacon guiding him closer to the core of existence.

His mind races, each thought branching into infinite possibilities. Memories of smelling fresh basil, the warmth of a sun-kissed tomato, the sizzle of olive oil in a pan–they all intertwine with his current pursuit. He maps these sensory experiences onto the geometric shapes, visualizing time folding in on itself and consciousness becoming a fluid entity. It's an alchemical process, a melding of past passions with newfound clarity.

As the minutes tick by, Jaxon becomes increasingly absorbed in his work. The studio, cluttered as it is, transforms into a sacred space where the boundaries between the physical and the metaphysical blur. The scent of incense grows stronger, intertwining with the fragrance of burning sage, creating an atmosphere thick with spiritual energy. The walls seem to close in, not with oppression, but with a sense of intimacy, as if the room itself is cocooning Jaxon in a protective embrace, urging him to delve deeper into the mysteries he's unraveling.

Hours blur into moments as Jaxon's confidant occasionally interjects with questions, each one a catalyst for deeper insight. "How do you know these shapes correspond to your experiences?"

"Because they resonate," Jaxon replies, his voice a mix of awe and certainty. "When I meditate or breathe in rhythm with them, it's like tuning into a higher frequency. My thoughts, they align in ways that make sense of the chaos."

As the first rays of dawn begin to filter through the grimy windows of the studio, Jaxon steps back from the board. The equation, a lattice of complex patterns and symbols, stands finished. It pulses with a latent energy, seeming to promise worlds beyond the known. He feels a profound connection to it, an understanding that transcends logic. This is not just a mathematical formula; it is a gateway to alternate dimensions of consciousness.

Jaxon's heart swells with a mixture of pride and trepidation. He knows what lies ahead is uncharted territory–both a potential for enlightenment and the risk of untethered minds. The early morning light casts a golden glow on the whiteboard, illuminating the fractal patterns with an almost mystical aura.

He stands in awe, grappling with the enormity of what he has created. This moment, a culmination of his fall and rise, is both a triumph and an omen. As the light grows stronger, Jaxon feels a shift within, a merging of his past and future selves. The equation, now birthed into the world, shimmers with the promise of infinite understanding, silently beckoning him and those who dare to explore its depths.

The studio, once a place of isolation, now feels like a portal to the infinite. The air hums with energy, and Jaxon knows that his life has changed irrevocably. The path ahead is both terrifying and exhilarating, a journey that will push the boundaries of his mind and soul. He takes a deep breath, steeling himself for what lies ahead.

Daylight streams through Jaxon's studio, casting beams of light over the cluttered expanse of his makeshift lab. The air is filled with an almost tangible buzz of anticipation as he sets up a series of experiments. His mind, a labyrinth of ideas both grounded and ethereal, is reflected in the meticulous yet chaotic arrangement of equipment around him. Fractal generators hum softly, their displays flickering with patterns that seem to pulse with life. Projection devices stand at attention, ready to visualize the geometric wonders Jaxon has theorized and experienced during his altered states.

Jaxon's movements are deliberate, each action infused with a sense of purpose. He places a projector on the tallest stack of books, ensuring it captures the entire surface of the projection screen. The screen soon flares to life, displaying a cascade of swirling, ever-evolving fractals. Each shape, each twist and turn of the patterns, is a manifestation of the universe to him—a glimpse into the fabric of reality.

Taking a step back, Jaxon focuses on his breath, drawing it in slowly, letting it expand within him. He begins a series of breathwork sessions, each breath a tether between his mind and the expansive equations scattered across the whiteboards. Every deep inhale and exhale synchronizes his physiological state with the rhythm of the fractals, allowing him to access deeper layers of consciousness. His body becomes a conduit, a bridge between the tangible world and the infinite possibilities of his mind.

As he slips deeper into his breathwork, Jaxon feels his awareness expand. The boundaries of his physical form begin to blur, merging with the patterns projected on the screen. His consciousness stretches outward, reaching toward the edges of the fractals, as if he could touch the very fabric of reality. The sensation is both exhilarating and terrifying, a reminder of the immense power he is tapping into.

Breathing deeply, Jaxon eases into a trance-like state. His body starts to move intuitively, each gesture a part of a larger, unseen dance. The geometry of his movements mirrors the equations he's created, his limbs tracing arcs and spirals that align with the intricate patterns appearing on the screen. Sweat beads on his brow as he loses himself in the dance, the physical exertion merging with the intellectual high of deciphering the cosmic tapestries he's only begun to understand.

Throughout this process, Jaxon's mind traverses an intricate network of thoughts and emotions. He recalls the days of his addiction, the haze that dulled his senses and numbed his genius. The memory fuels his fervor, propelling him to explore these higher states of awareness. The flashing equations on the whiteboard are not just symbols; they represent his journey from the abyss to enlightenment, each one a key unlocking another layer of reality.

Historical echoes fill his thoughts, spanning centuries of mathematicians and philosophers who used numbers to unravel the mysteries of existence. Pythagoras' sacred geometry, Euclid's elements, and the chaotic beauty of Mandelbrot's sets–all converge within Jaxon's consciousness. This lineage of knowledge makes him feel part of something vast and ancient, further amplifying his conviction that his fractal equation is not merely an academic triumph but a sacred relic–a blueprint of the cosmos. Pausing from the dance, Jaxon grabs a marker and returns to the whiteboard. His movements are fluid, almost frenetic, as he begins to scribble corrections and new insights, informed by the sensations coursing through his body and the epiphanies flashing in his mind. Each stroke of the marker feels electric, a tangible link between his physical self and the abstract realms he's navigating. He adjusts variables, redraws geometric shapes, and reconfigures the intricate latticework of his equation, recognizing how his thoughts, feelings, and bodily sensations are deeply interwoven into this grand tapestry of understanding.

The studio hums with potential energy; sunlight filtering through the windows bathes the room in a golden glow, illuminating the conduits of his experiments. The space itself feels alive, a sacred ground where art meets mathematics, creativity converges with science, and the individual intertwines with the universal.

Jaxon feels a profound connection to the cosmos, an intimacy with the universe that is as unsettling as it is exhilarating. Standing amidst his experiments, he senses that he is on the verge of something monumental–a revelation that could either uplift humanity or plunge it into chaos. The weight of this knowledge is palpable, pressing down on him even as it lifts his spirits with the promise of discovery.

He silently contemplates his late mentor, who once guided him through the labyrinthine intricacies of culinary art. The recollection brings a pang of loss but also a sense of responsibility. His mentor's belief in the transformative power of art propels him to reconcile his new mathematical insights with the beauty of creation. From beyond the veil of death, his mentor's lessons still echo, urging Jaxon to tread carefully.

As he stands there, amidst a sea of equations and fractals, Jaxon knows the path ahead is fraught with challenges–both intellectual and ethical. He sees the interwoven destinies of those around him: the seekers, the skeptics, and the lost. His studio, now a hallowed space of cosmic inquiry, resembles the crucible of his own soul, ready to test the boundaries of human understanding. The sunlight reflects off the whiteboard, making the equations shimmer like ancient runes.

Jaxon takes one last, deep breath. In this moment of clarity, he acknowledges the enormity of his task. Daylight bathes him in a warmth that suggests hope and renewal, as if the universe is offering its silent approval. The air pulsates with the electric possibility of what lies ahead–an uncharted realm of consciousness yet to be fully explored.

Engulfed in the intensity of his revelations, Jaxon readies himself for the next phase of this odyssey.

Jaxon stands at the center of his studio, its clutter transformed into an intimate gathering space. Soft light filters in from carefully placed lamps, casting a warm glow against the eclectic mix of art and scientific diagrams plastered on the walls. The air hums with anticipation as he welcomes an eclectic group–artists, spiritual seekers, and curious minds—all drawn by whispers of his newfound insights.

The studio, which once served as a place of solitary reflection, now buzzes with the energy of collective curiosity. The attendees, a motley crew of individuals drawn from various walks of life, fill the space with a sense of shared purpose. Some are artists, drawn by the promise of new inspiration; others are spiritual seekers, eager to explore the depths of consciousness that Jaxon has hinted at. All are united by a common desire to push the boundaries of understanding.

The murmur of introductions fades as Jaxon steps forward. His once haunted eyes now flicker with determination. "Thank you all for coming," he begins, his voice a blend of calm and fervor. "Tonight, I want to share something that has become the core of my existence—a fractal equation that bridges our consciousness with the deeper dimensions of the universe."

He gestures to a large whiteboard filled with swirling geometric shapes and complex equations. The attendees lean forward, some intrigued, others skeptical. Jaxon senses their mix of admiration and doubt, their need for understanding driving their curiosity.

"Imagine that each shape and curve on this board is not just a mathematical construct but a gateway," he explains, pointing to an intricate spiral. "These forms represent pathways to alternate dimensions of consciousness, accessible through intense focus, breathwork, and meditation."

He encourages the group to settle into a sitting position, closing their eyes, and guiding their breaths. "Let the equation guide your mind," Jaxon instructs softly, "but always stay connected to your breath. It's your anchor."

The room falls into a meditative silence, punctuated by the sounds of deep, rhythmic breathing. Some participants start to experience intense visuals—vibrant fractals bursting into existence behind closed eyelids, a symphony of colors and shapes that seem to sing of the universe's secrets. Jaxon watches them closely, knowing the fragility of this boundary they've crossed.

As minutes pass, the atmosphere grows palpable with a mixture of awe and strain. A few attendees achieve moments of profound clarity, their faces serene and radiant. One artist, tears streaming down her cheeks, whispers, "I see it... everything is connected," her voice trembling with revelation.

Others, less fortunate, begin to show signs of distress. A young man clenches his fists, his breathing becoming erratic as the complexity of the equation overwhelms him. Jaxon feels a knot tighten in his gut; he's seen this before–the fine line between enlightenment and madness.

Without breaking his calm, he raises his voice just enough to pierce their inner worlds. "Return to your breath. Ground yourself in your senses," he urges, his tone now imbued with urgency. The tension in the room thickens as some participants struggle to obey, their consciousness splintering under the weight of the abstract concepts.

One seeker begins to rock back and forth, muttering unintelligible fragments of the equation, eyes wide with panic. Jaxon approaches him, placing a steadying hand on his shoulder. "Breathe with me," he commands gently but firmly, synchronizing his rhythm with the man's.

The room, once filled with hopeful exploration, now thrums with a sense of impending crisis. Participants who had initially been focused and serene now waver, eyes opening to reveal confusion and fear. Jaxon moves among them, offering grounding words and gestures, pulling them back from the precipice of their minds.

As the last glimmers of daylight fade, the studio is bathed in the dim glow of its interior lights. Jaxon, sweating and slightly breathless, finally sees the attendees returning to a semblance of stability. The initial excitement has been replaced by a sobering realization–this knowledge holds power but also profound risk.

He guides the group to an end, his voice a thread of calm weaving through their heightened states. "Let's take a moment to ground ourselves," he says, encouraging them to touch the earth beneath them, feel the solidity of the floor.

The room gradually returns to a calmer state, the chaotic energy dissipating. Jaxon stands in the center, his face etched with both relief and the weight of responsibility. Eyes meet his, some reflecting gratitude, others veined with wariness. The collective experience has left them all altered, threaded with the threads of both discovery and danger.

The studio, now tinged with the aftermath of their shared expedition, feels like a battleground of the mind, a space where the infinite beauty of the cosmos wrestled with the delicate human psyche. As Jaxon ushers the last participant back to a state of calm, he understands more deeply than ever the gravity of the path he has forged.

The gathering, once envisioned as a celebration of discovery, has become a sobering reminder of the perils that accompany the pursuit of knowledge. The studio, once filled with the excitement of exploration, now feels heavy with the weight of what has been uncovered. Jaxon knows that he must tread carefully in the days to come, balancing the desire to push the boundaries of understanding with the responsibility to protect those who seek to follow him.

As the last of the attendees leave, Jaxon is left alone in the dimly lit studio, the weight of the evening's events pressing down on him. The air is thick with the residue of emotions–fear, awe, confusion– hanging in the space like an invisible fog. Jaxon stands amidst the remnants of the gathering, his thoughts swirling like the fractals he has spent so much time deciphering.

He moves slowly to the whiteboard, now filled with the complex, swirling patterns that have become both his obsession and his burden. The lines and curves seem to shift and dance under his gaze, revealing and concealing truths in a tantalizing game of hide-and-seek. Jaxon reaches out to touch the board, his fingers hovering just above the surface as if afraid to disturb the delicate balance of understanding he has achieved.

The studio, now silent and still, feels like a different place than it was just hours before. The energy of the gathering has dissipated, leaving behind a void that Jaxon can feel in the very air around him. The whiteboard, once a symbol of hope and discovery, now seems to carry a darker weight–a reminder of the responsibility that comes with unlocking the secrets of the universe.

Jaxon takes a deep breath, feeling the cool air fill his lungs, grounding him in the present moment. He knows that he cannot ignore the events of the evening, cannot brush aside the fear and confusion that some of the attendees experienced. The equation, so powerful and beautiful in its complexity, is not something to be taken lightly. It is a tool, a key, but it is also a weapon–capable of cutting through the fabric of reality and revealing the hidden layers beneath.

He steps back from the whiteboard, his mind racing with thoughts of what lies ahead. The road he has chosen is fraught with danger, not just for himself but for those who seek to follow him. The knowledge he has unlocked is powerful, but it is also perilous, and Jaxon knows that he must find a way to guide others through the labyrinth without losing them in the process.

As he stands alone in the dim light, the weight of responsibility settles on his shoulders like a heavy cloak. The equation, the fractals, the patterns–they are all part of something much larger than himself, something that he has only just begun to understand. The path ahead is uncertain, but Jaxon knows that he cannot turn back now. The journey has begun, and there is no going back.

With a final glance at the whiteboard, Jaxon turns and walks slowly out of the studio, the door creaking as it closes behind him. The night is cool and still, the air crisp with the promise of a new day. As he walks down the empty street, Jaxon feels the weight of the evening's events pressing down on him, but he also feels a flicker of hope. The road ahead is long and uncertain, but Jaxon knows that he is not alone. The universe is with him, guiding him, and he is ready to face whatever comes next.

Chapter 6

BROKEN SYMMETRY

The community center is a shadowy, hushed refuge amidst the constant din of the city, its low-lit interior cloaked in a veil of secrecy and anticipation. Jaxon Keller steps through the entrance, his heart pounding with a mixture of dread and curiosity. The flickering light from dusty chandeliers barely illuminates the room, casting eerie, elongated shadows over the faces of those gathered. They are all seekers, drawn by the gravitational pull of his fractal equation, each one seemingly chasing an elusive truth.

The room, with its high ceilings and dark wood paneling, is filled with a heavy sense of expectation. The scent of old books and worn leather permeates the air, mingling with the faint aroma of coffee brewing in a corner. The walls are lined with faded posters of past events–community outreach programs, yoga workshops, and art exhibitions–all of which seem out of place amidst the tension that now grips the space.

The air buzzes with low murmurs and sporadic laughter, underscored by the soft rustle of paper and the metallic clink of a teapot being poured. It's a fragile peace, teetering on the cusp of chaos. Jaxon feels the weight of their expectation pressing upon him, a tangible force that sharpens his awareness of the mind's delicate balance. He moves quietly to a seat near the center, his eyes flitting over the expectant faces encircling him, their expressions ranging from zealous fixation to quiet desperation.

As the meeting begins, the seekers take turns sharing their experiences with the equation. One by one, they recount their ventures into the labyrinthine cosmos of consciousness–each story blending incomprehensibly with the next. There's a frenetic energy in their words, as if they are attempting to grasp at shadows, trying in vain to verbalize the ineffable.

An older man named Tom speaks first, his voice wavering as he describes seeing "endless spirals" and hearing an "universal hum." Jaxon listens, his unease mounting. Tom's account begins coherently but soon dissolves into a cascade of disjointed phrases. Next is Sandra, a young woman with dark circles under her eyes. She grips her chair as she explains how her vision tunnels and fractals rearrange every time she closes her eyes. Her speech speeds up, words tumbling into one another in a hurried rush.

Jaxon's throat tightens. Each recounting is a grim mirror to his own early experiences. He catches sight of Sarah, sitting across the room. Her fingers fidget with the hem of her cardigan, eyes darting about desperately. Suddenly, she rises, rambling.

"Fragments," she says, voice pitched high, "it's all fragments! Broken pieces... eyes like portals..."

Jaxon's heart sinks as Sarah devolves into nonsensical phrases. Her distress is palpable, a contagion spreading through the room. The other seekers react, their nervous energy amplifying. Chairs scrape against the wooden floor, voices overlapping. The once gentle murmurs turn jagged, sharp. A man yells out from the back, "Make it stop! Just make it all stop!"

Jaxon stands, attempting to restore some semblance of order. "Please, everyone, calm down. We need to breathe, alright?"

His voice is a strained calm, trying to slice through the thickening fog of confusion. But it's too late. Panic ripples through the group, and Jaxon can almost see the fractures forming in their minds–splintering against the weight of these newfound realizations. The lights overhead flicker, a malfunction that adds to the disconnected, surreal atmosphere.

Jaxon's mind races. The room seems to pulse with a heartbeat of its own, an oppressive rhythm that presses against his ears. He watches Sarah's face twist in abject terror, her eyes glazed over. The scene around him distorts his perception, dreamlike, and he is haunted by the thought that his creation birthed this chaos. His heartbeats thunder in his ears, deafening him to the turmoil of voices.

He reflects bitterly on his past–the endless nights numbing his mind with substances. The way his own consciousness had teetered on the brink, fragile and precarious. He had intended enlightenment, fellowship with cosmic truths, but instead, he sees only fragmented minds and broken spirits. Jaxon's gaze locks on Sarah, whose mouth forms silent screams. His creation, his equation, was meant to open doors, but it has become a labyrinth with no exit.

Amid the chaos, memories of his attempts to escape his own pain rush back, intertwining with the reality before him. He feels the guilt of those nights, his brain dulled by chemicals, and realizes that the hope he clung to now torments others. His attempts to interject are suffocated by the overwhelming cacophony. Questions bombard him: Can he undo this? Has the fracture already spread too deep?

Jaxon steps closer to Sarah, grasping her shoulders gently but firmly. "Sarah, look at me," he says, his voice trembling with urgency. She doesn't respond, eyes vacant, lost in the depths of her mind.

Someone tries to leave, stumbling over chairs, and another shouts something incoherent about light and darkness. Jaxon watches in helpless despair as the meeting disintegrates. The seekers' mental states unravel before his eyes, a tapestry of human consciousness tearing apart at the seams. His vision blurs, tears threatening to spill as the gravity of his creation's impact bears down.

The community center, once a place of potential, is now a theater of madness. The haunting echoes of disjointed conversation reverberate in Jaxon's mind, each fragment a sharp jab in the heart. He stands amidst the wreckage of their minds, feeling his responsibility press down like a specter at his shoulder. As he looks around, he is left with a chilling realization: his creation enthralled them, yes, but it was also their unraveling. The weight of his unintended role in their undoing settles deep within him, marking a pivotal moment in his journey–a journey that now demands atonement.

The chaos of the room builds to a fever pitch, and Jaxon feels his control slipping through his fingers. The air is thick with the scent of fear, the sound of desperate voices clashing in a discordant symphony. He is overwhelmed by the sheer magnitude of the situation, the knowledge that his creation, once a beacon of hope, has now become a source of despair.

With a final, desperate effort, Jaxon tries to regain control. "Please, everyone, stop!" he shouts, his voice breaking through the cacophony. "We need to take a step back, breathe, and find our center."

But his words are lost in the din. The seekers are too far gone, their minds shattered by the weight of the equation. Jaxon watches helplessly as the room descends into madness, the very thing he had sought to prevent now unfolding before his eyes.

As the last vestiges of order disintegrate, Jaxon is left standing in the eye of the storm, surrounded by the broken pieces of his creation. The realization hits him with the force of a physical blow: he has failed. The equation, which he had believed would bring enlightenment, has instead brought chaos and suffering.

He takes a deep breath, the weight of his guilt pressing down on him like a physical force. He knows that he must find a way to make this right, to undo the damage he has caused. But as he looks around the room, at the shattered minds and broken spirits, he wonders if it is already too late.

Jaxon shuts the door of his cluttered apartment and leans against it, the dim light filtering through grimy windows casting long shadows on the walls. Towers of paper, remnants of dishes, and mathematical notes scattered haphazardly across every surface tell the tale of a man consumed by a relentless pursuit. The echoes of the frantic meeting still resonate in his mind, each participant's face a ghost that lingers in his thoughts.

He drags a chair to the center of the room, its legs scraping against the wooden floor, and sits heavily. The room is his sanctuary and his prison, each object a fragment of his fractured past and uncertain present. The walls, once a blank canvas during his days as a renowned chef, are now plastered with complex equations and geometric patterns, the tangible remnants of his descent into a different kind of obsession.

As he flips through his notes, his mind drifts back to the initial spark of inspiration that led to the creation of the fractal equation. It was meant to be a beacon of enlightenment, a bridge to higher consciousness. But now, he questions whether it was crafted to captivate minds or merely to dazzle them with its complexity.

Memories of the seekers' faces intrude, their initially eager expressions morphing into confusion and terror. He sees Sarah, her eyes wide with fear, her words dissolving into gibberish. The cacophony of voices in the community center rises in his mind, each overlaying the other, forming an incoherent symphony of madness. He cannot ignore the stark realization that his work–the equation–has led to their unraveling.

A profound sense of guilt begins to swell within him, like a wave gathering strength before the inevitable crash. He recalls Mark's shattered psyche and the terror in his eyes as he spoke of "the void." These seekers, drawn to the promise of transcendence, have instead been ensnared by an abyss of their own making, deeper and more treacherous than he had ever anticipated.

Jaxon stands and paces the room, running a hand through his tangled hair. The light outside dims further, the city sinking into twilight. His apartment, once a space filled with the vibrant energy of culinary creation, now feels stifling, heavy with the weight of his newfound knowledge. The heady scent of forgotten spices mingles with the acrid sting of old papers, creating an atmosphere that clings to him, suffocating and inescapable.

He returns to his desk, staring at the notes scattered before him—each equation, diagram, and hastily scribbled thought a step towards his current predicament. The promise of understanding, of unlocking the mysteries of consciousness and reality, lies entangled within these pages. Yet, the cost has become painfully apparent.

His thoughts turn inward, reflecting on the broader implications of his creation. The fractal equation was supposed to illuminate the hidden dimensions of the mind, offering glimpses into realms of thought and existence beyond the ordinary. But its very nature–its complexity and power—demands a level of cognitive and spiritual maturity that few possess. Those who falter face a mental fragmentation that mirrors the equation's infinite complexity, a cruel reflection of their own inner chaos.

The realization hits him hard: his work does not just explore the boundaries of human cognition; it pushes them, sometimes beyond repair. The seekers, in their quest for enlightenment, have become victims of their own unprepared minds. Society, often dismissive of those who teeter on the edge of mental stability, now views them as shattered remnants, casualties of an experiment gone horribly wrong. Jaxon cannot shake the image of the seekers as both explorers and casualties, their minds stretched thin across the boundaries of human understanding.

Determined and desperate, Jaxon resolves to delve deeper into the phenomenon. He must understand the true nature of the equation and find a way to mitigate its devastating effects. The burden of responsibility weighs heavily on him, a constant reminder of the fine line between genius and madness.

Jaxon settles at his cluttered desk, the dim light from a single lamp casting deep shadows as he scours his notes, looking for answers. His hands tremble slightly, yet his resolve hardens. This journey, fraught with uncertainty and peril, is one he must take. He owes it to the seekers, to those who have suffered, and to his own fractured soul.

As his eyes scan the pages, determination sets in, mingled with dread. The path ahead is fraught with danger, both to himself and to those who dare to follow. But now, with the weight of guilt pressing down on him, Jaxon knows he must move forward. The seekers' faces, their voices a haunting chorus, propel him into action. He will find the answers, whatever the cost. Hours pass in a blur as Jaxon immerses himself in his work, the apartment fading into the background as he dives deeper into the complexities of the equation. His mind races, connecting dots and unraveling the mysteries hidden within the intricate patterns. The night stretches on, the only sound the scratching of his pen on paper and the occasional creak of the building settling around him.

The more he delves into the equation, the more he begins to see the flaws in his approach. He had been so focused on the potential for enlightenment that he had overlooked the dangers, the inherent instability of the human mind when confronted with the vastness of the cosmos. He realizes now that the equation is not just a tool for exploration–it is a weapon, one that must be wielded with care.

But how can he mitigate the damage already done? How can he guide the seekers back from the brink, help them find their way out of the labyrinth he has unwittingly led them into? The questions swirl in his mind, each one more daunting than the last.

Finally, as the first light of dawn begins to filter through the window, Jaxon leans back in his chair, exhaustion weighing heavily on him. He has made progress, but there is still so much more to do. He knows that he cannot rest until he has found a way to control the chaos, to guide the seekers safely through the realms they have entered.

But for now, he allows himself a moment of respite. He closes his eyes, letting the weight of his guilt and responsibility wash over him. He knows that the road ahead is long and treacherous, but he is determined to see it through.

Jaxon fumbles with his phone, hands trembling, as the frantic voice on the other end slices through his already frazzled nerves. His friend's words are barely coherent, but the urgency is unmistakable. Mark has gone off the rails. The weight of the news hits Jaxon like a sledgehammer to the chest. Without another thought, he dashes out of his apartment, the remnants of his fractal equation scattered like cryptic breadcrumbs across his cluttered desk.

The public park looms ahead, its serene beauty now an ironic backdrop to the chaos within. Jaxon spots Mark immediately, a lone figure thrashing near the fountain's edge. The scenery melts into surreal distortions, nature's vibrant greens and blues clashing violently with the darkness enveloping Mark's mind. Mark's eyes are wild, a turbulent sea of fear and confusion. He babbles incessantly about the void, his voice a discordant melody that reverberates through the air.

Mark grips his head, his hair tangled and damp with sweat. "It's everywhere, Jaxon! I can see it all! The void, it's swallowing me!"

Jaxon approaches cautiously, his own breath quickening. He extends a hand, trying to ground Mark in the physical world.

"Mark, listen to me. It's Jaxon. You need to calm down. Breathe with me, okay? In and out."

But Mark cannot hear him over the cacophony of his own unraveling thoughts. "No! You don't understand! The equation... it's unlocked something, something we were never meant to see!"

Jaxon's heart aches as he watches his friend spiral further into madness. He searches desperately for a sliver of the Mark he once knew–a man filled with hope and curiosity. "Remember what we talked about? The patterns, the balance. Try to focus on something tangible. Look at the fountain. The water is real." Mark's eyes briefly dart to the fountain, but just as quickly as they find fleeting calm, they glaze over again. "The void is in the water too! It's everywhere, Jaxon! I can't escape it!"

Jaxon fights against his rising panic, speaking more urgently now. "Mark, stay with me. We hopped on this path together, remember? There's balance in everything, even in chaos."

Mark's laughter is hollow, echoing through the park. "Balance? There's no balance in what you've shown us. Only madness. Only the void."

Jaxon's mind races. He's grasping at the remnants of their shared history, moments of camaraderie and scholarly pursuit. He clings to the hope that these memories can anchor Mark, pull him back from the edge. But Mark is slipping too far too quickly. With a sinking heart, Jaxon realizes he cannot do this alone. His own enlightenment, born from the same equation, now feels like a double-edged sword. Each insight comes with the price of another's sanity.

He fumbles for his phone again, dialing emergency services. His voice cracks as he speaks to the dispatcher, his words stained with the weight of responsibility. "I need help. My friend is having a mental breakdown in Prospect Park, nearly rthe main fountain. Please, hurry."

The minutes stretch into lifetimes as they wait, Mark continuing to oscillate between frantic declarations and eerie silence. Jaxon kneels beside him, a silent guardian, his heart pounding with every nonsensical utterance Mark spills into the night. The shadows of guilt and helplessness loom large, choking him with their oppressive presence.

He remembers how Mark was one of the seekers, an eager mind drawn to the promise of the unknown. Their shared journey had begun with enthusiasm, but it had veered into perilous territory, leaving Jaxon to grapple with the consequences of his creation. The scene in the park is a harrowing testament to the fine line between genius and madness, a line they had crossed together.

Sirens wail in the distance, a mournful prelude to the arrival of help. As paramedics descend upon the scene, Jaxon steps back, feeling the crushing weight of failure settle over him. He watches as they tend to Mark, their practiced hands moving with confident urgency, performing the motions of salvation. But in his chest, Jaxon feels the hollow thud of failure; no paramedic could mend the fragmented consciousness that the equation had splintered.

The paramedics lift Mark onto a stretcher, securing him with straps that seem far too inadequate to contain the depths of his unraveling psyche. Jaxon follows as they wheel him to the ambulance, every step pulling at his soul. He wants to reach out, to soothe Mark's disjointed fears, but he knows that words hold no power here. Not anymore.

The ambulance doors slam shut with a finality that echoes in Jaxon's mind. He stands in the park, now eerily quiet, feeling the night press in on him. The world around him seems to slow, each heartbeat amplifying the silence. He is left alone with the weight of his actions, the emptiness of the park mirroring the void Mark so vividly feared.

As the ambulance disappears into the distance, Jaxon remains rooted in place, staring after it, feeling the cruel twist of fate. He had sought knowledge and found it, yet what he uncovered was a chasm too vast for even the brightest minds to navigate. In this moment, he grasps the true scale of the tragedy he has wrought.

The night deepens, shadows of trees stretching like spectral fingers. Jaxon takes a deep breath, tasting the cold bitterness of regret. His mind is a spiral of conflict, an endless loop of questioning every decision that led to this point. He feels neither genius nor enlightened, only a man bearing the unbearable burden of unintended consequences.

Steeling himself, he turns away from the park, the quiet resolve setting in. The night air is thick with the scent of earth and leaves, grounding him as he finds his path amidst the uncertainty. This cannot be the end. He must find a way to understand, to control the chaos his equation has unleashed. It is a responsibility he can no longer avoid.

Jaxon takes a final look at the now silent park, a place that has become a crucible of his guilt and determination. He walks away, every step echoing the vow to find a solution, to make right what has gone so horribly wrong. The night closes around him, but within that darkness, a flicker of resolve ignites, guiding him back to his desk, back to the work that waits, and to the answers that he must find, for Mark and for all the souls teetering on the edge of the void

Jaxon stands in front of the mirror in his bathroom, the dim light casting long shadows across his gaunt features. His eyes, once sparkling with the fervor of culinary creativity, now reflect the haunted depths of a man lost within a labyrinth of his own making. The seeker's faces flash before him–Sarah's wide eyes brimming with incoherent terror, Mark's haunted expression as he cried out about "the void." Every line on his face feels like an etching of the chaos he's unleashed. He's bare-chested, standing alone in the small, cluttered bathroom, surrounded by relics of a life that feels lifetimes away.

He grips the edges of the sink, the cold porcelain grounding him in the present moment. His thoughts swirl like the fractal patterns he now perceives the world in–complex, interwoven, infinite. The mirror offers no solace, only a stark reflection of the man he has become and the monstrous beauty of his creation. Images of his past life intrude–a bustling kitchen, dishes that danced with molecular gastronomy, a younger, confident Jaxon with a mind that saw flavors like a painter sees colors. The fall from grace, marked by addiction and isolation, seems both distant and immediate.

His breath fogs the glass, and with a trembling hand, he wipes it clear, revealing the eyes that have borne witness to unspeakable realms. In a sudden, painful clarity, he understands that the fractal equation, a portal to boundless dimensions, had fractured not just reality but the fragile psyches of those who dared engage with it. Each seeker was a mirror, reflecting the labyrinth of his own consciousness and the chaotic beauty of the universe. Their minds, unprepared for the vastness they encountered, splintered like glass under pressure.

A wave of guilt envelops him, tightening around his chest, making it hard to breathe. He can see them—Mark's desperate pleas for understanding, Sarah's fragmented sentences that echoed with madness. His mind becomes a storm of sorrow and regret, each lightning bolt a flash of their distorted faces. He recalls the night of the lightning strike, the agony and rebirth it brought, the moment that steered his journey from culinary genius to cosmic conduit. Every step since then led to this—standing before a mirror, grappling with the ruin he has wrought.

His reflection blurs as unshed tears sting his eyes. The bathroom's silence amplifies the eerie cacophony of his thoughts. He speaks aloud, a hollow whisper, "What have I done?" The words linger in the stale air, melding with the memories that haunt him. Anguish ekes into his voice, a mix of fear and remorse. He feels the weight of his genius turned curse and the lives fractured by his quest for understanding.

Then, amidst the swirling storm of his mind, a moment of searing clarity. The seekers' fragmented states, their descent into madness, are not merely tragic consequences; they are mirrors reflecting his own fractured reality. The equation doesn't merely expose new dimensions—it amplifies what lies within. Jaxon, a man torn by addiction and brilliance, had unwittingly crafted a cosmic mirror. The seekers approached it with their inner chaos, and it returned their reflection in the form of madness.

This realization pierces him, leaving behind a raw wound of vulnerability. He sees, all too clearly, that his responsibility extends beyond intellectual curiosity. He is a harbinger of fractured truths, a creator of chaos, and a reluctant guardian of a gate that leads to untamed realms. His torn reflection stares back, unflinching, challenging him to face the consequences of his creation without denial or evasion.

Slowly, resolutely, determination replaces despair. The refracted light of the universe has long danced in his mind, illuminating paths only he can navigate. He must find a way to mitigate the harm, to transform the equation from a weapon of chaos into a tool of careful enlightenment. He envisions a path forward—a way to help seekers prepare their minds, to shield them from the overwhelming dimensions he has unlocked.

The mirror becomes an oracle, revealing not just his flaws and failures but the potential for redemption. Jaxon leans closer, his reflection splitting into an infinite number of perspectives. His eyes, once brimming with despair, now hold a flicker of indomitable will. For every realm he sees, there is a route to understanding, for every fracture, a path to healing.

He takes a deep breath, feeling the cold air fill his lungs with each deliberate inhale. Looking into the depths of his eyes, Jaxon vows to find a solution. The equation cannot be erased, but its impact can be altered. He straightens up, his reflection merging with his resolve, blending the brilliance of his mind with the empathy of his heart. Determination surges through him–where there is chaos, there can be order; where there is madness, there can be clarity.

The last vestiges of his former self, the celebrated chef who turned chaos into culinary art, now forge a new path. Jaxon will transmute his cosmic insights into a force for stability and understanding. He steps back from the mirror, the decision made. Tonight, he becomes not just a seeker of truths but a guardian of minds. He exits the bathroom, the weight of his resolve matching the gravitational pull of the cosmos he has glimpsed. The night's silence is punctuated by a newfound purpose–vindicating those he unwittingly led into chaos and steering his knowledge toward a more benevolent horizon.

The road ahead is long and uncertain, but Jaxon knows that he is ready to face whatever challenges lie ahead. He has been given a second chance, a chance to make amends and to guide others safely through the labyrinth of consciousness he has helped to create. And he will not squander it.

With a final glance at the mirror, Jaxon turns and walks out of the bathroom, ready to begin his journey of redemption.

Chapter 7

PATTERNS OF REDEMPTION

Bright flashes of news segments ripple across television screens, each one dissecting Jaxon Keller's extraordinary transformation and his confounding equation. Talking heads, brimming with enthusiasm or skepticism, debate the implications, their voices blending into a background hum that saturates the atmosphere. Jaxon's name echoes through living rooms, coffee shops, and academic halls, igniting a collective fascination with this enigmatic figure who now stands at the crossroads of science and mysticism.

Crowds of journalists and curious onlookers have gathered outside Jaxon's modest home, a practical embodiment of his desire for simplicity amid a life once marked by culinary excess. His house, now surrounded by a sea of bodies and cameras, feels like a fortress under siege. The fragrant scent of fresh cedar mingles incongruously with the tangy metallic smell of restless anticipation, pervading the air. In their eagerness, they inadvertently trample the modest flower bed that Angelica had planted, adding an unspoken urgency to the scene. The encapsulated scent of turned earth and crushed petals only heightens Jaxon's sense of invasion.

Inside, Jaxon paces like a caged animal. His home, once a sanctuary of reflective quietude, has been transformed into an echo chamber of anxiety. The pressures weigh down on him like a material entity, pressing against his chest and constricting his breath. He tries to focus, but the cacophony outside–raised voices, the incessant shutter clicks of cameras–shatters any semblance of peace. Thoughts roil within his mind, an uncomfortable juxtaposition of his sacred realizations and the chaotic demands of the world outside. Beneath the surface of his newfound enlightenment festers a deep-rooted fear: the fear of being misunderstood, of his sacred equation being diluted into sensational snapshots for a ravenous public.

The walls of his home, once a sanctuary, now seem to close in on him. Every creak of the floorboards, every distant shout from the reporters outside, amplifies his sense of confinement. The once comforting clutter of his notes and sketches now feels oppressive, the symbols and patterns a constant reminder of the burden he carries. Each fractal he once admired for its beauty now taunts him with its complexity, its implications sprawling out into realms he can barely comprehend, let alone control.

Craving isolation, Jaxon decides to make an escape out the back door. The backyard, usually a private haven where he meditated amongst the rustling of trees and the whispers of the wind, now feels like a precarious tightrope. He moves stealthily, his movements a careful dance of evasion. As he steps onto the gravel path leading to the side alley, every crunch beneath his feet speaks his presence.

Before he can reach the gate, an enthusiastic reporter–an ambitious young woman with wide, eager eyes–captures his path. "Mr. Keller, just a few questions!" she implores, her voice breaking the still afternoon air.

Cornered, Jaxon feels his heart rate spike. He takes a moment, gathering the tangle of thoughts into coherence, before reluctantly agreeing to a brief interview. He knows this is a moment that demands careful articulation, yet words seem to slip through his grasp like grains of sand.

"What inspired your fractal equation?" she asks, pressing a microphone toward him, her curiosity bright but unsophisticated.

Jaxon takes a breath, feeling the skepticism wrap around his thoughts like a vice. "It wasn't a singular inspiration," he begins, his voice carrying the weight of worlds. "This equation... it's a convergence of countless dimensions of understanding. It transcends traditional mathematics and veers into the realm of consciousness itself."

The reporter's brow furrows slightly, a silent plea for simplification evident in her expression. Sensing her perplexity, he feels a wave of frustration. How can one distill the infinite into palatable sound bites? He continues, struggling to thread the needle between revelation and simplicity. "Imagine if every thought, every feeling, and every moment existed on an infinite fractal spectrum. This equation maps that, it ties the material to the ethereal, the known to the unknown."

She nods, more in hope than understanding, and fires another question. "What do you say to those who call you a modern-day prophet?"

He shrugs off the title with a mixture of irritation and fear. "I'm not a prophet," his voice edges with vulnerability. "I'm a man who brushed against the fabric of the cosmos and brought back a thread. It's not about prophecy—it's about exploration, about opening doors our minds didn't even know existed."

The interview concludes, leaving Jaxon feeling exposed and tethered to an unyielding spotlight. He retreats back into his home, closing the door against the outside world, but he can still feel the probing stares pressing against the walls. His sanctuary has become a stage, every corner imbued with the expectation of an audience. Jaxon sinks into a chair by the window, staring into the now tattered garden, and feels a sharp pang of loss. He meditates on the irony; enlightenment was never meant to be a performance, and yet here he is, a reluctant actor in a spectacle far larger than himself.

The pressure mounts, seeping into every aspect of his existence. Every interview, every murmur of curiosity, chips away at his resolve, amplifying his insecurities and ethical dilemmas. Is he guiding humanity towards an awakening, or leading them into an abyss? The lines blur in the quietude of his thoughts, the cacophony of the external world blending with the chaos within.

He thinks back to simpler times, when his world was confined to the kitchen, where the only expectation was to create something delicious, something that brought joy. Now, the stakes are infinitely higher, and the joy he once found in creation has been replaced by a gnawing anxiety. The world demands answers from him, explanations for the mysteries he's uncovered, but he knows that some mysteries aren't meant to be solved, only experienced. How can he convey this to a world that craves certainty, that fears the unknown?

The philosopher-king who once commanded a kitchen now faces a conundrum that no recipe can solve. His equation is a child of the cosmos, yet the world demands it be simplified, understood, possessed. The weight of this realization presses down on him, a cosmic burden he never asked for. Alone in his cluttered thoughts, Jaxon reflects on the fervent noise surrounding him, a noise that promises to only grow louder.

He whispers to the solitary plant by the window, "What have I done?" The silent leaves waver gently, as if the universe itself offers an enigmatic response, one he may never fully comprehend.

The local community center feels like a buzzing nexus of anticipation, its usual gymnasium space now transformed into an informal arena for the public demonstration. Rows of folding chairs are packed with faces reflecting a spectrum of curiosity and skepticism, their conversations an undercurrent to the tension in the room. Projectors hum softly, ready to bring mathematical marvels to life. A makeshift stage is set at one end, where Jaxon Keller, caught in the intense spotlight, wrestles with the weight of expectation.

He feels the collective gaze of the audience–academics armed with notebooks, esoteric seekers clutching crystals, and everyday people drawn by the novelty of the event. The community center's fluorescent lights bounce off the shiny surfaces, casting geometric patterns that already seem to blur with the fractals about to be displayed. Among them sits Dr. Miles Carter, his skeptical gaze piercing through the crowd, and Maya Santos, her eyes both supportive and questioning, a complex cocktail of admiration and silent challenge.

Jaxon's fingers tremble slightly as he adjusts his microphone, aware that the room's ambient hum grows quieter, a vacuum of expectancy. He begins by projecting fractal patterns onto the screen; each intricate design, vibrant and pulsating, seems to breathe as if alive. The audience gasps collectively, enraptured by the beauty of the visuals, reminiscent of organic blooms and cosmic structures. However, beneath the surface, the beautiful patterns provide a stark reminder of the equation's complexity–a universe reduced to mathematical elegance.

As he speaks, his voice wavering, Jaxon attempts to navigate the chasm between abstract concepts and everyday understanding. "This equation is not just a mathematical trick... It's a map of dimensions, an insight into the fabric of our consciousness and, by extension, reality itself." He sees nods from some corners, but furrowed brows from others. His explanations are laden with terms like 'multiverse', 'quantum entanglement', and 'sacred geometry'. Each word reverberates in the air, either sparking wonder or deepening confusion.

Dr. Miles Carter rises from his seat. "Mr. Keller, can you provide empirical validation for these claims? Without rigorous testing, what separates your equation from mere conjecture?" His voice is a controlled blend of curiosity and disbelief, echoing the broader scientific community's demand for evidence. The room tightens with collectively held breath, waiting for Jaxon's response.

Jaxon's mind races. He stammers slightly, "The empirical... it's challenging to test something that transcends our current scientific tools. This is a blend of experience and mathematics. It bridges what we perceive with what truly exists."

Carter's eyes narrow. "So, it's unverifiable by conventional scientific standards? Isn't that problematic?"

The audience murmurs, a maelstrom of whispers. Doubt surfaces, like a dark tide eroding the initial wonder. Jaxon feels his pulse spike, a surge of anxiety threatening to choke out his words. He takes a deep breath, pushes forward. "Empirical science has its limits. This equation opens doors to realms beyond those confines. Those who engage with it profoundly–through meditation, through altered states–they can experience..."

Before he can finish, another voice interrupts—a skeptic hidden behind thick-rimmed glasses. "Sounds like pseudoscience. How can we trust a drug-addict-turned-philosopher over centuries of scientific methodology?"

Jaxon's face flushes. The reminder of his past is a dagger twisting in the wound of self-doubt. He feels the ground tilting, his confidence slipping like sand through fingers. Maya catches his eye, her look urgent, a silent encouragement to hold firm. He grips the edge of the podium, steadying his resolve. "I understand the skepticism. But this isn't just about science; it's about consciousness, tapping into the profound. It's a tool for exploring depths we've only dreamed of."

Despite his best efforts, the presentation grows progressively shakier, the room's mood fracturing into pockets of support and dissent. Some nod in eager belief, eyes glistening with hope, while others shake their heads, anchored in their skepticism. The tension becomes palpable, a weight that presses down on Jaxon's chest, making it harder to breathe, to think clearly.

As he wraps up, he whispers, "Thank you for listening," though his gratitude feels hollow, muffled by exhaustion. The applause is scattered, tepid. The lights dim, and the fractal patterns fade to darkness, leaving only the stark, cold reality of the community center.

Jaxon stumbles off the stage, head spinning, heart heavy. The mixture of conflicting reactions feels like a physical force pressing against him. He sees Carter scribbling in his notebook, more questions undoubtedly forming, while Maya approaches with a look of concern and reluctant admiration. But it's all too much. He needs air, space to breathe, away from the cacophony of judgment.

Stepping out into the dimly lit corridor, Jaxon lets the cool air wash over him, a temporary balm for his frayed nerves. The overwhelming sense of unease lingers, wrapping around his thoughts like a tight band. In this strained moment, he wonders if he's earned the right to wield such knowledge, sensing the burden of the enlightenment he inadvertently unleashed.

The corridor's dim lighting feels like a refuge, a buffer between the harsh judgment inside and the world outside. But as he leans against the cool wall, trying to center himself, the weight of it all crashes down. The applause, the questions, the accusations—it's all a blur, a cacophony that drowns out his thoughts. His breath comes in short gasps, his vision narrowing as the pressure builds.

He thinks back to the day he first began to see the patterns, the fractals, the glimpses of other dimensions. Back then, it was pure—an exhilarating discovery that filled him with wonder. But now, it feels tainted, sullied by the world's demands for proof, for validation. They don't want to understand—they want to dissect, to categorize, to strip it of its mystery until there's nothing left but cold, hard facts.

But the equation was never meant to be dissected. It was meant to be experienced, to be felt. How can he make them understand that? How can he make them see that some things are beyond empirical validation, that some truths exist in the spaces between the lines, in the whispers of the universe that can only be heard if you listen with your soul?

But as he stands there, trying to catch his breath, he realizes that it's not just them—it's him too. He's doubting himself, questioning the very thing that brought him to this point. The skepticism, the disbelief—it's seeping into him, poisoning his thoughts, turning his wonder into fear.

He pushes off the wall, forcing himself to move, to walk, to do anything to break free from the spiral of doubt. The corridor stretches out before him, long and empty, the lights flickering as if mocking his uncertainty. He knows he should go back, face the crowd, answer their questions. But right now, he just needs to get away, to find a moment of clarity amidst the chaos.

The exit looms ahead, a door that promises escape, if only for a little while. He quickens his pace, his footsteps echoing in the empty corridor, each step a desperate attempt to outrun the doubts that chase him. He pushes open the door and steps outside, into the cool night air that feels like a balm against his heated skin.

The city sprawls out before him, a sea of lights and noise that feels overwhelming and comforting at the same time. He walks down the steps, his mind still racing, still trying to make sense of everything that's happened. But the more he thinks, the more the doubts grow, until they're a tangled mess of fears and uncertainties that he can't seem to untangle.

He finds a bench at the edge of the park and sits down, his head in his hands as he tries to quiet the storm inside. The night is calm, the air cool and crisp, but his mind is anything but. The equation, the skeptics, the applause—it all swirls together in a chaotic dance that he can't control.

He knows he's at a crossroads, a point where he has to make a choice. Does he continue down this path, even though it's filled with doubt and uncertainty, or does he turn back, retreat into the safety of the known, the comfortable? But deep down, he knows there's no turning back. The equation, the patterns—they're a part of him now, a part of his soul that he can't ignore, no matter how much he might want to.

But he also knows he can't do it alone. He needs help, guidance, someone who can see beyond the skepticism, beyond the doubt, someone who can help him navigate this new world he's found himself in. He thinks of Maya, of her quiet support, her belief in him, and he knows she's the one who can help him find his way.

But first, he needs to find his own strength, his own belief in what he's discovered. He needs to remember why he started this journey in the first place, the wonder, the excitement, the sheer joy of discovery that filled him when he first saw the patterns. He needs to hold onto that, to let it guide him through the darkness, through the doubt, until he can see the light again.

He takes a deep breath, letting the cool air fill his lungs, and slowly, the chaos in his mind begins to settle. The doubts are still there, the fear still lingers, but they're not as overwhelming as they were. He knows the path ahead won't be easy, but he also knows it's the path he's meant to take.

With a newfound sense of resolve, he stands up, ready to face whatever comes next. The city is still alive around him, the lights twinkling in the darkness, the noise a distant hum that no longer feels so overwhelming. He starts to walk, his steps steady and sure, as he heads back to the community center, ready to face the crowd, ready to share his discovery with the world, no matter what they might think.

The café is a haven of chatter and the aroma of brewing coffee. Jaxon walks in, seeking solace among familiar faces, hoping to find a moment of peace. Sunlight filters through the large windows, casting patterns on the wooden floors and tables scattered with cups, saucers, and discarded napkins. Conversations ebb and flow around him, a symphony of intellectual exchange that used to comfort him but now feels alien.

Jaxon joins a table with friends, but the sense of camaraderie feels stretched thin. His once lively demeanor is now overshadowed by a growing sense of isolation. As he sips his coffee, the vibrant, rich taste fails to bring him the usual pleasure. His mind, accustomed to creating transcendent dishes, now grapples with the complexities of an equation that has thrust him into an uncomfortable spotlight.

He overhears two professors at a nearby table, their conversation seeping into his awareness. One of them, a stout man with graying hair, waves a hand dismissively. "The whole thing sounds like pseudo-science," he says, his tone dripping with skepticism. "Where's the empirical validation? It's just another sensational story."

The other professor, a younger woman with bright eyes and a bemused expression, leans forward. "But can't you see the potential? It's not just about validation. It's about expanding our understanding of consciousness. The patterns he describes–if real, they could revolutionize our field."

Jaxon's heart quickens. He can't help but feel a surge of determination to engage. He edges closer and clears his throat. "Excuse me," he begins, his voice trembling slightly. "I'm Jaxon Keller."

The stout professor eyes him critically. "Yes, we know who you are."

Jaxon swallows hard, trying to find the right words. "My equation... it's not just a story. It's a way to interpret dimensions of consciousness that we've never considered before."

The younger professor's eyes light up. "Tell us more. How did you derive it?"

Jaxon begins explaining, but his thoughts, usually fluid and clear, stumble over one another. The professors exchange glances, and the older man interrupts. "Look, Mr. Keller, without rigorous testing and peer review, your equation remains speculative at best."

The conversation shifts back to their own dialogue, dismissing Jaxon's input as if he were an irrelevant participant in a larger game. Their self-assured tones, laden with academic jargon, highlight their intellectual elitism. They chew on their perspectives like they're savoring fine wine, oblivious to Jaxon's struggle for validation.

Feeling alienated, Jaxon's eyes dart around the café. The once-inviting space now seems oppressive. He stands abruptly, mumbling a hasty goodbye to his friends and heads to the door, pushing through the heavy air of dismissal that clings to him.

Stepping outside, he's met by the cool night air, a sharp contrast to the café's warmth. The streetlights cast a tired glow over the pavement, and the distant hum of the city fades into a backdrop of obscured thoughts. Each step feels heavier as he grapples with the dual perceptions of his work–celebrated by some, ridiculed by others.

His thoughts swirl like chaotic fractals, each branching into an infinite maze of doubt and uncertainty. Jaxon wonders if his equation is a beacon of insight or merely a magnet for ridicule and misunderstanding. The weight of his creation bears down on him, amplifying his sense of isolation. The clarity he seeks seems as elusive as ever, hidden behind a veil of skepticism and divided interpretations.

Standing under the dim streetlight, Jaxon feels the growing distance between his intention and the world's reception. The loneliness of misunderstood genius, estranged from both the intellectual community and his own former life as a chef, digs deeper into his core. The once-inspiring connections of flavors and cosmic patterns now seem to slip through his fingers like sand. He inhales deeply, trying to find grounding amidst the turmoil.

Jaxon's mind drifts back to his past, recalling the applause of diners who once marveled at his culinary masterpieces. Back then, his art was tangible, universally appreciated without the burden of proving its validity through equations. That life, though, was tinged with addiction and escape. Now, his pursuit of enlightenment is shadowed by a different kind of struggle–a quest for meaningful connections that respect and understand his new insights.

But the clarity that once defined his work as a chef has been replaced by the murky waters of philosophical and mathematical exploration. The lines between art and science, between creation and destruction, blur more with each passing day. Jaxon feels like a man standing on the edge of a vast chasm, peering into an abyss that offers both enlightenment and madness, unsure which one will claim him.

As he turns to walk away from the café, the weight of misunderstanding hangs heavy on his shoulders. Jaxon contemplates the journey ahead, where every step seems to sink deeper into the quagmire of public perception and personal doubt. The irony is not lost on him; the man who once crafted dishes to bring people together now wrestles with an equation that has driven a wedge between himself and the world.

The street is quiet, the usual bustle of the city muffled by the late hour. Jaxon walks aimlessly, his thoughts a whirlpool of uncertainty. The clarity he once found in the kitchen, in the precise art of culinary creation, now feels like a distant memory, a faint echo of a simpler time. The equation, once a source of pride, now feels like a burden, a weight he's not sure he can carry.

But as he walks, something shifts within him. The doubts and fears that have plagued him begin to recede, replaced by a steely resolve. He knows that the path ahead won't be easy, but he also knows that he can't turn back. The equation is a part of him now, as much a part of his identity as his love for cooking once was. It's a journey he's destined to take, no matter the cost.

Feeling the sting of solitude, Jaxon continues down the darkened street, the distant glow of the café fading behind him. He embraces the uncertainty like an old, reluctant friend, knowing that his path is a lone one, guided by the faint light of understanding that flickers within his mind.

Jaxon hunches over his laptop, the harsh glow of the screen contrasting with the dim lighting of his home office. His inbox overflows with unread emails, each subject line more sensational than the last. "Exclusive Interview Offer!" "Join Our Podcast!" "Collaborate on Your Equation!" The superficial nature of the requests feels like an assault on his mind. He grips the edge of the table, knuckles whitening. The equation, once a blossoming flower in his subconscious, now seems to be wilting under the weight of external interpretation.

Jaxon clicks on a few messages, scanning through accolades peppered with requests for simplification, as if his work were a mere product to be marketed. His heartbeat quickens, each message adding another layer of anxiety. He recalls his days as a chef, where complexity was celebrated, not stripped down to its bare bones. The irony isn't lost on him: he sought to create something unfathomably profound, and yet he now stands at the precipice of its trivialization. His hands shake slightly as he closes the laptop, painfully aware that shutting off the screen won't silence his buzzing thoughts.

The doorbell chimes, a discordant interruption amid his reverie. Jaxon straightens, wiping his palms on his jeans before opening the door to find two members of an esoteric group standing there. Their flowing robes and serene smiles are at odds with the aggressive intent in their eyes. They introduce themselves with names that sound as if plucked from ancient texts, offering elaborate greetings laced with promises of enlightenment. Stepping aside, Jaxon allows them in, his reluctance masked by a veneer of politeness.

The meeting room is dimly lit, shadows dancing along the walls as the group takes their seats. Incense wafts through the air, meant to be calming but only succeeding in making Jaxon's pulse race faster. The leader of the group, a man named Lysander, begins to speak in a tone that skirts the edge of reverence and coercion. "Brother Jaxon, we believe your equation holds the key to unlocking higher planes of consciousness. Aligning it with our spiritual practices will elevate its impact."

Jaxon shifts uncomfortably in his seat. The room feels claustrophobic, the fabric of his reality seeming to close in around him. "My equation isn't meant for spiritual practices," Jaxon responds, his voice hoarse. "It's a mathematical representation of a reality we barely understand."

Lysander's eyes glitter with unholy eagerness. "But you cannot deny the equation's spiritual resonance. It's our duty to guide humanity toward the enlightenment it offers."

Panic claws at Jaxon's insides. He remembers the people who had suffered mentally from attempting to understand the equation without adequate preparation. He pictures their fragmented minds, their eyes reflecting a chaos that should never be unleashed. "The equation is dangerous," he stammers. "Without proper cognitive maturity, it can–"

"–Lead to transcendence or tribulation," Lysander interrupts smoothly. "With our guidance, it will be the former." His words are a velvet hammer, unyielding beneath the surface of genteel insistence.

Jaxon feels his frustration boiling over into anger. "You're turning something complex into a tool for your own agenda. That isn't what it was created for!" His voice rises, cracking under the strain of remaining composed. He rises abruptly, his chair scraping against the floor, a harsh noise slicing through the oppressive atmosphere.

Lysander's serene mask slips for a moment, a glimmer of frustration evident. "Brother Jaxon, you must understand.."

But Jaxon doesn't wait to hear the rest. He storms out, the heavy door slamming shut behind him. Outside, the fresh air is a balm, but his mind churns like a storm-tossed sea. He walks aimlessly through the city streets, the cacophony of car horns and distant chatter doing little to quell his inner turmoil. The streetlights blur into trails of light as he moves, each step heavy with the realization of how easily his work could spiral out of control.

As he walks, he reflects on his past–his addiction, the brief moments of clarity washed away by a tide of self-destruction, and the bolt of lightning that had reformed his mind. The path to enlightenment was not straight; it twisted and turned, resembling the fractal patterns he now saw in everything. But such knowledge came with the price of immense responsibility–a weight he fears he cannot bear.

Isolated despite the city's throng, Jaxon contemplates the danger of his equation in unworthy hands. It's an alchemical creation, potent and volatile, capable of both profound insight and catastrophic madness. Under the streetlight's flicker, he reflects on the chasm between his intentions and the world's interpretations, each step resonating with the weight of potential misuse. His desire for personal redemption feels dwarfed by the monumental responsibility of safeguarding a fragile, transformative insight.

Jaxon knows he must protect the sanctity of his creation. Yet, as shadows stretch longer and night envelops the city, he doubts whether he can preserve its integrity against the relentless tide of misunderstanding and misuse. Walking further into the night, his steps echo with an increasing desperation to reconcile his fragmented life and the fragile brilliance of the equation–uncertain of what lies ahead.

The weight of his own thoughts threatens to crush him. Each step feels heavier than the last, the burden of his creation pressing down on him with an almost physical force. He stops beneath a streetlight, the faint glow illuminating his haggard face, and wonders how it all came to this. How did a simple desire for understanding spiral into something so complex, so dangerous?

The equation, once a source of pride and joy, now feels like a curse. It has brought him nothing but confusion and chaos, alienating him from the very world he sought to understand. He remembers the seekers, their faces twisted in fear and confusion, and wonders if he's done more harm than good. Is the knowledge he's uncovered worth the pain it has caused?

He looks up at the night sky, the stars barely visible through the haze of city lights. They twinkle faintly, distant and cold, indifferent to the struggles of the man below. Jaxon feels a profound sense of isolation, as if he's the only one who truly understands the weight of the knowledge he carries. The world around him feels foreign, a place he no longer belongs.

But despite the doubts and fears that plague him, there's a part of him that refuses to give up. He knows that the equation holds the key to something profound, something that could change the way humanity understands reality. But he also knows that it's a double-edged sword, capable of cutting both ways. The challenge lies in wielding it wisely, in finding a way to share it with the world without causing harm.

He thinks of Maya, her quiet strength and unwavering support. She's been with him through it all, never wavering in her belief in him. He knows he needs her now more than ever, needs her guidance to navigate the treacherous path ahead. But even as he thinks of her, he feels a pang of guilt. He's dragged her into his world of chaos and uncertainty, and he wonders if it's fair to ask her to stay.

But he also knows that he can't do it alone. He needs her, needs her insight and her strength, needs her to help him find a way forward. The path ahead is uncertain, but with her by his side, he feels a glimmer of hope, a faint light in the darkness.

As he continues walking, his steps more determined, he resolves to find a way to protect his creation, to ensure that it's used for good, not harm. The equation is his responsibility, and he won't let it fall into the wrong hands. He'll find a way to share its knowledge with the world, but on his terms, in a way that preserves its integrity.

The night is quiet as he walks, the city slowly coming to life around him as the first light of dawn begins to break through the horizon. The world is still, as if holding its breath, waiting for what comes next. And for the first time in a long while, Jaxon feels a sense of calm. The path ahead is still fraught with danger, but he's ready to face it, ready to take on the challenge.

He walks through the quiet streets, the city slowly waking up around him, and for the first time in a long while, he feels a sense of purpose. The equation, the knowledge it holds, is his responsibility, and he's ready to protect it, to guide it, to ensure that it's used for the greater good. The night may be dark, but the dawn is coming, and with it, the promise of a new beginning.

Jaxon walks on, the first rays of sunlight beginning to warm the city, and feels a renewed sense of hope. The journey is just beginning, and he's ready to face whatever comes next, with Maya by his side, and the weight of the world on his shoulders.

But as he walks, he knows that the road ahead will not be easy. There will be challenges, there will be doubts, but he's ready to face them all. For the first time in a long while, he feels like he's on the right path, a path that will lead him to the answers he's been searching for, a path that will lead him to the truth. And with that, he walks on, ready to face whatever comes next.

Chapter 8

BENEATH THE SURFACE OF STARS

Jaxon steps into the dimly lit café, a sanctuary for seekers of eclectic wisdom and intellectual refuge. The walls are adorned with vibrant art–abstract shapes and colors that seem to twist and turn like the thoughts in his mind. Shelves filled with worn books line the corners, and each table is a little island of contemplation, occupied by those lost in philosophical musings or whispered debates.

He spots her instantly–Maya, a woman with a commanding presence, deeply engaged in conversation with a small group. The soft lamplight casts a warm glow on her expressive face as she talks passionately about consciousness. Her words, though initially lost in the café's ambient noise, draw him in like a magnet.

He approaches, curiosity piqued, the recent notoriety of his fractal equation still an unfamiliar weight on his shoulders. He feels a mix of pride and unease, remnants of his struggle with addiction nibbling at his confidence. The equation, though groundbreaking, is a double-edged sword, a conduit for enlightenment or madness.

"Mind if I join the discussion?" Jaxon asks, his voice cutting through the hum of the café.

Maya's eyes meet his, a flicker of recognition followed by intrigue. "Of course," she says, gesturing to an empty seat.

Jaxon settles in, the eclectic décor—a blend of mismatched chairs and tables set against murals of cosmic scenes—adding to the surreal ambiance. He's reminded of past philosophical debates, moments when his mind raced with ideas yet felt the crushing weight of insecurity. This is different; the stakes are higher.

"What are your views on the nature of reality and consciousness?" he asks, diving straight in.

Maya's eyes light up. "I believe consciousness is a collective experience, shaped and molded by our shared interactions and emotions. It's not just an individual phenomenon but a tapestry woven from all our lived experiences."

He leans forward, captivated by her conviction but ready to challenge. "But isn't individual consciousness paramount? Each person's perception is unique, and it's this individuality that defines our reality," he counters.

Their discussion intensifies, voices rising and falling like waves. Maya's hands move animatedly as she speaks, her belief in the interconnectedness of human experience firm. "Imagine consciousness as a vast ocean," she says, "each human merely a wave contributing to its depth and complexity. Our emotions, thoughts, and actions ripple out, influencing and shaping the collective."

Jaxon crosses his arms, a reflective gesture that speaks to his internal struggle with the concept. "But each wave is distinct, Maya," he insists. "Our individual consciousness is what brings innovation and change. Without it, we're just part of a homogenous whole, losing the very essence that makes us human."

Their conversation ignites the air around them, as if the café's very walls are absorbing and reflecting their intellectual fervor. The eclectic art seems to swirl and twist, mirroring the energy between them. Jaxon can't help but feel a spark of admiration for Maya's passionate arguments, even as he wrestles with his defensiveness.

He recalls past debates, moments when his confidence wavered under scrutiny. This time, however, there's a thrill in the exchange, a sense that each point made is a step toward deeper understanding. Maya's collective view of consciousness clashes dynamically with his own individualistic perspective, fueling a competitive yet respectful dialogue.

The café, with its dim lighting and eclectic décor, becomes a microcosm of their vibrant exchange. The scent of strong coffee mingles with the earthy aroma of burning incense, creating an environment that challenges the mind while soothing the spirit. The chatter of other patrons fades into the background, leaving Jaxon and Maya in their own intellectual bubble.

As the discussion reaches a fever pitch, frustration begins to surface. Both are entrenched in their beliefs, yet visibly intrigued by the other's perspective. Maya's eyes flash with determination, while Jaxon's brow furrows in concentration.

The conversation is a dance of ideas, each step bringing them closer yet highlighting their philosophical divide. Despite the growing intensity, there's an undeniable magnetic pull between them, a mutual curiosity that keeps the debate from devolving into mere argument.

By the end, both Jaxon and Maya are visibly frustrated but deeply intrigued. There's a tangible energy between them, an unspoken agreement that their conversation is far from over. They part with a charged silence, leaving the eclectic atmosphere of the café to drift into the cool night air.

As Jaxon walks away, his mind buzzes with Maya's words. He feels a shift within, a burgeoning desire to explore not just the equation's implications but the profound ideas it has brought into his life. The café's remaining patrons continue their own conversations, unaware of the seeds of partnership that have been sown in the dimly lit corner.

The encounter leaves him both challenged and exhilarated, setting the stage for what promises to be an intellectually and emotionally transformative journey.

Jaxon leans back into his chair, the scent of freshly brewed coffee mingling with the ambient hum of conversation and the low notes of a jazz quartet playing softly in the corner. The café, with its dim lighting and eclectic decor, feels like a portal to another era–one where ideas flow freely and minds connect over shared thoughts and passions. Bookshelves filled with worn volumes line the walls, and vibrant artwork provides a colorful backdrop to the steadily deepening bond between Jaxon and Maya.

Jaxon sips his coffee, feeling the warmth spread through him, a counterbalance to the electric tension simmering between them. Maya leans forward, her eyes alight with the spark of intellectual curiosity. "How can you be certain," she challenges, "that individual consciousness is paramount? Isn't our sense of self deeply intertwined with others?"

Her question worms its way into Jaxon's thoughts, poking at the fragile edifice of his ego. He sets down his cup, its ceramic clink a momentary punctuation. "Because," he counters, "the universe itself is experienced subjectively. One's perception shapes reality. Without the individual, the collective is nothing more than shadows on a cave wall."

Maya nods slightly, acknowledging his point as she stirs her tea. "True," she concedes, "but isn't it the shared narratives and emotions that give meaning to those individual experiences? Our connections mold us, define us."

The debate weaves intricate patterns through the air, a dance of words and ideas. Jaxon's mind flickers to the past, the lonely nights lost in a haze of substances, his self-designed prison of isolation. These memories stoke the embers of self-doubt; he wrestles with the fear that his newfound clarity might be nothing more than ephemeral. Jaxon's gaze narrows, not in anger but in contemplation. "I once thought like that," he admits. "But then, I lost myself—not in the collective, but in the void within. Only by returning to my individual core did I start to find my way back."

Maya's face softens momentarily, revealing a flicker of shared pain. She takes a slow breath before speaking. "I understand loss intimately. My best friend... addiction claimed him. It was through our shared journey—one that ultimately led me to philosophy—that I began to grasp the interconnectedness of our paths. His struggle was not his alone but a reflection of the world around him."

Her admission touches something deep within Jaxon, an acknowledgment of suffering that bridges the gap between them. Both have wandered through their own labyrinths of despair, and this common ground begins to shift the nature of their interaction. Jaxon's stern facade crumbles slightly. "Perhaps," he says gently, "we're two sides of the same coin. Individual and collective, separate yet united in the grand tapestry."

The café's intimate nook cocooning them, filled with the murmur of distant conversations, seems to amplify the quiet intensity between them. Maya asks, "Have you ever tried to reconcile these worlds through direct experience? Meditation, perhaps?"

Jaxon considers her words, the suggestion both a challenge and an invitation. "I have," he admits, "but always alone. The idea of a group meditation session is... intriguing."

Maya's eyes light up with earnestness and something more—an unspoken understanding. "Then let's do it together. Experience might illuminate truths that philosophy alone cannot."

The decision made, they look around the cozy café, now a backdrop to their budding camaraderie. They find an upcoming group session on the café's bulletin board—an invitation to a shared journey. The anticipation hums between them like an unseen current, each contemplating the potential revelations and the subtle bond forming between them.

The session begins with silence, their breaths synchronizing in the dim light. As their minds quiet, Jaxon finds himself slipping into a meditative state, the equations and fragments of his cosmic understanding merging with Maya's shared presence. Time folds, the boundaries of self blur. He senses Maya's thoughts intertwining with his, her life force becoming a thread in the fabric of his reality.

Post-meditation, they linger over coffee once more, newfound insights glistening like dew on their consciousness. Maya speaks first. "That was... profound. I felt your presence, your journey, as if it were my own."

Jaxon nods, grasping her hand with a newfound sense of connection. "And I yours. It seems our philosophies, while divergent, can find common ground in shared experience."

Their gazes lock, a silent exchange filled with mutual respect. They have traversed beyond mere words, stepping into a realm where intellect and spirit converge. In that moment, they agree silently to explore these new dimensions together, not as rivals, but as companions on a shared path.

They sit back, sharing a knowing glance. This connection–rooted in challenge and bound by understanding–is the start of something profound, a promise of what lies ahead.

Jaxon stands at the threshold of his studio, the dim light from the hanging lantern casting shadows that dance with anticipation. He gestures for Maya to enter, a subtle nod signaling the start of their shared journey. The space is a blend of past and present–walls adorned with abstract canvases, remnants of his culinary triumphs, and new artworks reflecting his current obsession with the cosmos. The air buzzes with an electric charge, an invisible thread of excitement weaving through the room.

Jaxon moves to a small table near the center, meticulously arranging the elements for their DMT experience. He places the vial of the potent substance, a pair of comfortable cushions, and a woven tapestry depicting sacred geometry. His hands tremble slightly–a mix of excitement and fear. He glances up, meeting Maya's eyes. "Ready?"

Maya nods, her expression a mirror of his own mix of anticipation and apprehension. She takes a seat on one of the cushions, folding her legs beneath her. "Let's set our intentions," she suggests softly.

As they sit opposite each other, Jaxon's mind drifts to the complexity of their journey. His fall from grace, the lightning strike that realigned his neural pathways, and now this–an attempt to delve deeper, not alone, but with someone who challenges and understands him. He breathes deeply, centering himself, and speaks. "I intend to explore the interconnectedness of all things and seek understanding beyond my current grasp."

Maya follows, her voice calm and steady. "I intend to find clarity and wisdom, and to connect with the deeper layers of consciousness that bind us together."

Their intentions hang in the air, an unspoken pact, binding them to the experience about to unfold. Jaxon hands her the vial, and she takes it with a steady hand. As they place the substance beneath their tongues, the room begins to shift–walls bending, colors morphing, and sounds blending into an otherworldly symphony.

As the DMT takes hold, reality fractures into a kaleidoscope of impossible patterns. Geometry dances in front of their eyes– triangles, fractals, spirals–all spinning into a sacred cosmic dance.

They float through dimensions where light curves and bends in defiance of physics, where time becomes a fluid entity, folding in on itself like ripples in a pond. Shared visions unfold, vivid and surreal.

Jaxon sees himself not just as an individual but as a node in a vast, glittering web of consciousness. Each thought branches into infinity, creating a new universe with every blink. Colors meld, tastes emit sounds, and the scent of memories wafts through invisible winds. Maya's presence is a constant—her form is fluid yet distinct in the shifting reality. They share visions of ancient civilizations, sacred rituals, and cosmic symphonies—mutual echoes of their philosophies brought to life.

Jaxon feels the boundaries between them dissolve, their minds touching and intertwining in ways words could never describe. He sees Maya's inner world—the traumas, the hopes, the relentless pursuit of meaning. She, in turn, experiences his chaotic blend of addiction, genius, and the relentless drive for redemption. The fractal equation they've debated appears before them in radiant, pulsating energy; it's not just seen but felt, a visceral truth stitched into the fabric of their consciousness.

Gradually, the intensity wanes, and they find themselves back in the studio, the cacophony of the universe quieting back into the familiar hum of their surroundings. Their breaths synchronize, both coming back to the tangible world, forever changed.

Maya breaks the silence, her voice a whisper of awe. "I saw... everything. The patterns, the truths we only theorized about—they're real."

Jaxon nods, struggling to put his own revelations into words. "It's like our consciousness is a mirror, fractal in nature, reflecting infinite layers of reality. Each argument we had... they weren't contradictions. They were facets of a singular truth."

They sit in silence, the depth of their shared experience settling in. Jaxon glances around his studio, the once mundane objects now infused with a new significance. Each painting, each relic of his past, hints at a greater web of existence. He reaches out, taking Maya's hand, his touch grounding them both.

"This changes everything," he says quietly, the weight of his words pressing down on them gently.

Maya squeezes his hand, nodding. "We need to be careful. This knowledge... it's powerful but fragile."

Their eyes meet, and in that silent exchange, they understand the depth of their bond. Words are unnecessary; their shared vision has bridged gaps and created a connection that transcends the physical realm.

As twilight descends, they sit quietly, reflecting on the magnitude of what they've experienced. The future is uncertain, but their shared journey has just begun, a path illuminated by the fractal light of their newfound understanding.

As twilight descends upon the tranquil park, the sky is painted with hues of indigo and auburn. The air is crisp, carrying the scent of autumn leaves and the distant murmur of the city. Jaxon and Maya find themselves on a weathered bench near a large oak tree, its branches swaying gently in the evening breeze. The intensity of their earlier DMT experience lingers, wrapping them in a shared quietude that feels almost sacred.

"I'm grateful you chose to walk this path with me," Jaxon begins, his voice laden with sincerity. The deep lines etched into his face soften, and his eyes, usually shadowed by doubt, reveal a glimmer of hope. "Your perspective...it's like a mirror reflecting parts of me I've avoided for so long."

Maya, nestled in the soft fabric of her shawl, turns to him with a warm yet penetrating gaze. "I see it too," she replies, her tone both gentle and firm. The evening light catches the intricate patterns of her tattoos, casting delicate shadows on her skin, as if the sacred symbols themselves are part of the conversation. "We mirror each other's struggles and strengths, Jaxon. Our pasts, though painful, anchor us to a deeper understanding of what it means to truly awaken."

Jaxon nods, a weight lifting from his shoulders, barely discernible but undeniably present. "For so long, I drowned myself in substances, thinking I could escape the pain of losing my mentor, my purpose, my own self-worth," he confesses, his voice almost a whisper. "The lightning strike–it was like the universe startling me awake, forcing me to see the fractures within."

The park around them grows quieter as the last of the evening joggers and picnickers disperse, leaving a serene stillness. Maya's eyes soften further, empathy coursing through her. "I understand more than you know," she says, a faint quiver in her voice. "I too have danced with addiction, tried to fill the void left by my friend's death with substances and distractions. But it's through our shared pain that we find the true essence of enlightenment."

As their words settle between them, a deeper connection roots itself in the fertile soil of their vulnerabilities. They both realize that their quests for knowledge are not isolated; instead, they are intertwined paths through a shared landscape of suffering and revelation.

Jaxon's thoughts turn inward, reflecting on the ethical heaviness of his discovery. "The equation," he murmurs, almost as if to himself, "it's more than just a series of mathematical principles. It's a window into the fabric of consciousness, capable of profound understanding but equally capable of causing harm."

Maya contemplates his words, her brow furrowed in thought. "The knowledge you've unearthed," she says slowly, "is like a double-edged sword. We must tread carefully, respecting its power while guiding others in humility and wisdom. Misused, it could indeed fracture minds as it illuminates them."

The moon rises, casting a soft glow over the clearing. Jaxon feels the gravity of their conversation, the moral weight pressing yet also enlightening. "We must ensure that those who seek this knowledge are ready," he asserts. "That their minds and souls are not burdened with chaos, lest they become distorted reflections of their inner turmoil."

Maya's expression brightens with resolve. "This shared responsibility—I'm willing to help navigate it with you. My pursuit of enlightenment is not just personal; it's communal. We owe it to ourselves and to those touched by your equation to guide them with care."

The night deepens around them, a blanket of stars emerging in the sky, symbolizing infinite possibilities. Jaxon extends his hand towards Maya, a gesture laden with hope and trust. "Let's make a pact," he says, his voice steady. "To support each other, to share this burden and this journey. To ensure that our path is one of integrity and mutual respect."

Maya takes his hand firmly, her touch a blend of warmth and determination. "We will walk this path together," she agrees. "Through the light and the shadows, standing as both seekers and guides."

Their bond solidifies in that moment, a silent agreement echoing through the quiet park. They sit in companionable silence, the weight of their pact settling comfortably, as if destiny itself has woven their threads together. As they look out over the park, the twilight deepening into night, a shared sense of purpose and unity blossoms between them, heralding the dawn of a powerful alliance.

Chapter 9

A TASTE OF TRANSCENDENCE

Dr. Miles Carter leans back in his office chair, eyes glazed over the email displayed on his laptop screen. His colleague's words blur before him, each sentence questioning the authenticity of Jaxon's equation. Frustration knots his muscles, and irritation simmers just beneath his professional demeanor. The sterile, dimly lit room, filled with the hum of fluorescent lights and stacks of academic journals, feels stifling. Carter's mind races back to his formative years, those long days and nights spent buried in textbooks and peer-reviewed papers, the relentless pressure to conform to established paradigms beating like a drum in his ears.

A notification pings, pulling him back to the present. The seminar in the lecture hall is about to begin. He exhales deeply, the weight of the email lingering as he gathers his notes and heads down the corridor.

The lecture hall is a stark contrast to his office–bright, spacious, and filled with murmurs of conversation among colleagues. The walls are adorned with portraits of celebrated scientists, their eyes seemingly judging the proceedings. Dr. Carter takes a seat in the middle row, clutching his notepad. As the seminar begins, the atmosphere turns tense, a palpable skepticism hanging in the air.

The presenter, a distinguished figure in the scientific community, takes the podium and starts dismantling Jaxon's work, labeling it pseudoscience with a dismissive wave of his hand.

"Jaxon's equation," the presenter declares, "is nothing more than a mathematical hallucination—a fantastical creation devoid of empirical substantiation."

Carter's grip tightens on his pen. The words cut deep, reminding him of the rigid academic environment that has molded his career. The seminar room, filled with rows of earnest faces, each belonging to a colleague vying for recognition, suddenly feels oppressive. He recalls his own upbringing, the constant drive to validate conventional scientific truths, dreams often crushed by the weight of tradition.

Unable to remain silent, Carter stands, his voice cutting through the murmur of agreement. "You dismiss Jaxon's equation too easily. What if it holds truths that the conventional paradigms fail to address? Aren't we, as scientists, supposed to explore the unknown?"

A prominent scientist, Dr. Elaine Foster, rises from the back, her eyes cold and calculating. "We are bound by the scientific method, Dr. Carter. Venturing into unstudied, mystical ideas without concrete evidence is reckless."

The hall grows hushed, every gaze fixed on them. Carter feels the weight of collective scrutiny, the pressure to conform clashing with an insistent spark of curiosity. His pulse quickens, his voice steadies, "Yet, all great discoveries stemmed from questioning the accepted norms. Perhaps Jaxon's equation challenges us to expand our understanding."

The debate intensifies, drawing the attention of other attendees who edge their seats closer. Tensions flare as arguments spike back and forth, the room's atmosphere electrified with intellectual fervor. The symbolic portraits on the walls seem to shift, their stares now reflecting a mix of disapproval and intrigue.

Leaving the seminar, Carter returns to his office, his mind adrift in conflicting thoughts. The skepticism of his peers gnaws at him, a stark reminder of the academic battles he's faced. His desk, cluttered with printouts and notes, serves as both a comfort and a burden. A single photo frame catches his eye—his best friend from his teenage years, a brilliant mind lost to the chaos of mental illness. The memory of that loss fuels his determination as Dr. Carter opens his laptop. Shadows lengthen and the room darkens, but his resolve hardens.

He types feverishly, drafting questions, contemplating approaches. His curiosity, once stifled by the weight of academic expectations, now finds a flicker of freedom. Carter closes his laptop, determination etched into every line on his face. He is committed to uncovering the truth behind Jaxon's equation, despite the backlash and skepticism from his peers.

In the quiet of the office, only the rhythmic ticking of a wall clock fills the air. Dr. Carter's resolve swells, certain of one thing—he must follow this path, regardless of where it leads.

In the dimly lit confines of his private laboratory, Dr. Miles Carter stands before a stark, oversized printout of Jaxon's equation, pinned with precision against the wall. The equation's intricate beauty casts a spell over him, its swirling symmetries a dance of numbers and symbols that defy conventional comprehension. The glow from the monitors provides the only source of light, casting long, unsettling shadows.

His mind, ever the fortress of rationality, now grapples with a growing unease. The formula beckons him to peer beyond the boundaries of established science. Carter's upbringing in a world steeped in empirical evidence and rigorous methodology had always made him skeptical of such uncharted waters. Memories flicker of long nights spent immersed in equations, reinforced by the stern yet supportive voices of mentors who praised logic over speculative wonder.

Taking a deep breath, he moves to his desk, where the hum of the computer soothes his restive nerves. His hands are deft and methodical as he inputs the equation into the system, selecting parameters to run simulations. Screens light up with a cascade of data and graphs. The lab becomes a sea of numbers, pulsating with raw, unfathomable potential. Carter's fingers hover over the keyboard before pressing 'Enter', sending the simulations into motion.

As the first sets of results stream in, they appear orderly, adhering to his expectations. But as the simulation progresses, the data begins to diverge, morphing into patterns that defy his linear understanding. The graphs twist and swirl, each curve a manifestation of some higher complexity that eludes his grasp. Carter's intrigue heightens, blending with a burgeoning sense of dread. Could these results be hinting at a dimension beyond the empirical, a frontier where the logical world dissolves into chaotic beauty?

Carter's mind reels as he watches the unpredictable outcomes unfolding on the screen. Every result poses new, unanswerable questions. The boundaries of reality as defined by his scientific principles begin to blur, eroding the bedrock of certainty that has long anchored his understanding of the universe. He feels the weight of this new reality pressing down on him, unsettling the very foundations of his beliefs.

His internal conflict sharpens, a knife's edge balancing curiosity and fear. The teachings of old mentors echo in his mind. "Stick to what can be measured," they had always urged, their voices a chorus of caution. Yet here he stands, seduced by the intangible, the equation's mysteries laying siege to his guarded fortress of rationality.

The fluorescent lights above flicker, coinciding with his own moment of doubt. Staring at the equation now printed and pinned, a chill runs down his spine. The realization that these symbols might reveal a truth too vast, too complex for the human mind to safely navigate seizes him. He mutters to himself, "What are you doing, Miles?" His voice, usually steady, quivers in the silence.

Carter sinks into his chair, rubbing his temples, the gesture an attempt to soothe the chaos swirling within his mind. His thoughts spiral, ranging from the exhilarating possibility of a breakthrough to the terrifying prospect of uncovering truths that could upend the very fabric of scientific understanding. Is it hubris to pursue this knowledge? Will it strip away the layers of reality until nothing familiar remains?

His relationship with Jaxon floats to the surface of his consciousness. The chef-turned-mystic had always viewed knowledge as an experiential journey, contrasting sharply with Carter's empirical approach. Their conversations often underscored this dichotomy, a dance of admiration and tension, skepticism and wonder. Jaxon, in his enigmatic way, spoke of realms beyond perception, of truths felt rather than calculated. Carter, ever the scientist, struggled to reconcile Jaxon's esoteric musings with his own disciplined inquiry.

Leaning back, staring into the enigmatic depth of Jaxon's equation, Carter feels his world shift. Waves of excitement and fear wash over him in alternating surges. The equation is not just an intellectual puzzle; it's an invitation–a portal to a new realm of understanding that challenges every tenet of his upbringing.

As the simulation's final results flutter on the screen like cryptic messages from an alien world, Carter sits back in his chair, enveloped by the numinous glow. His mind is a storm of conflicting thoughts, anticipation interwoven with trepidation. He knows decisions must be made. Paths must be chosen. The truth, whatever it may be, is waiting to be uncovered. But at what cost?

He stares at the enigmatic numbers, feeling the prickling sensation of standing on the edge of an abyss, where every answer only leads to more profound and troubling questions.

The café is bathed in the soft glow of dim lighting, its corners cluttered with forgotten memories collecting dust. The air carries the faint, poignant mingling of stale coffee and cigarette smoke, a common refuge for souls weighed down by burdens few can articulate. Most patrons blend into the background, their conversations muffled and indistinct, creating a hushed atmosphere where stories of desperation can breathe freely.

Dr. Miles Carter sits at a corner table, an island in this sea of whispered coherence. He's hunched over a small cup of coffee, methodically stirring its contents, waiting for the seeker. His eyes dart to the door each time it creaks open, a symphony of anxiety playing beneath his stoic façade. Finally, a figure enters–gaunt, disheveled–a living testament to the unraveling effect of Jaxon's equation.

The seeker takes a seat opposite Carter, hands trembling as they struggle to clutch the flimsy menu. Their eyes are haunted, fragmented prisms of shattered realities. "The lights," the seeker begins, voice quivering like a fragile thread, "they bend, twist into... other worlds."

Carter leans forward, catching every word. "Tell me about these worlds," he urges, his tone both steady and comforting.

Struggling to find coherence, the former seeker describes a DMT experience that transcends the bounds of comprehension. Their voice is a fractured mosaic. "Colors... they sang. I was more... less... a whisper in the void, and the void... whispered back." They pause, unsure if their words make sense even to themselves. "Jaxon's... equation... it's... it's a door, but... what's on the other side... it's not meant... for us."

The chaos of their sentences—half thoughts and aborted phrases—speaks volumes. Carter takes detailed notes, capturing the essence of their disjointed accounts with growing concern. Each erratic gesture, every tremor in their hands, is meticulously documented. His eyes narrow in deepening worry as he feels the weight of their plight.

Carter's mind races, piecing together the puzzle with each erratic anecdote. He senses the urgency beneath their stories, an emotional pulse that cannot be ignored. The equation, he realizes, is more than an abstract concept; it has real, tangible effects on those who dare to engage with it. People like this seeker, buried under layers of trauma and fragmented realities.

The café's dim light casts shadows across the seeker's weary face as they recount delusional episodes and fragmented realms. "I saw... my own death... and my birth, in... the same breath." Their breathing hitches, recalling the horror. "I was there, and... not."

Carter's pen scratches against paper, capturing each jumbled fragment. His growing sense of responsibility tightens around him like a vice. He listens, not just to the words, but to the desperate undercurrent that drives them, understanding the seekers' need to articulate the chaos within their minds.

The interview reaches a natural conclusion as the seeker's fragmented narrative trails off into silence. Carter closes his notebook, the weight of newfound responsibility pressing upon his shoulders. He understands that the seekers trust him to unravel their chaotic experiences, to validate the turmoil Jaxon's equation has stirred within them.

Stepping out of the café, Carter finds himself enveloped in the cool night air, his breath visible in the moonlit haze. He sits behind the wheel of his car, resting his head against the steering wheel for a moment, allowing the reality of what he has just heard to settle in. The murmured stories of the seekers replay in his mind, vivid apparitions that refuse to fade.

He begins to connect the dots, contemplating the seekers' accounts and the complexity of Jaxon's creation. Each testimonial, each tremor in their voice, hints at the dark consequences of engaging with the equation. These encounters are more than isolated incidents; they form a disconcerting pattern that threatens to unravel what he thought he knew about the boundaries of science and reality.

Carter starts the car, the engine's hum a stark contrast to the silence of his thoughts. As he drives, the implications of Jaxon's equation weigh heavily on his mind, a tumultuous storm brewing just beneath the surface. The night stretches out before him, an endless canvas marked with the pinpricks of distant stars, reflecting the vast unknown he navigates.

Back in his lab, Dr. Miles Carter hunches over his workspace, the dim glow of screens casting sharp shadows. The air is thick with anticipation, a tangible manifestation of his growing obsession. Staring at the intertwining glyphs of Jaxon's equation sprawled across the largest monitor, he feels an odd mix of awe and trepidation. The crisp, mechanical whir of the lab equipment contrasts sharply with the almost ethereal nature of what he's attempting to decipher.

Carter's fingers hover over the keyboard, almost afraid to disturb the mystical hieroglyphs birthed from Jaxon's enigmatic mind. His limited interactions with Jaxon had already revealed a man who saw the universe in a lattice of possibilities, a stark contrast to Carter's empirical upbringing. He inputs complex algorithms, his heart beating a tad too fast, wondering if his meticulous calculations could unlock the secrets Jaxon seemed to hold so effortlessly. The computer hums in protest as it processes the simulation, sweat forming a cold line down his back.

The screen bursts into a frenzy of fractals and swirls, data showering him in a cascade of incomprehensible beauty. Each result is more unfathomable than the last. Patterns within patterns; colors that shouldn't exist in any scientific spectrum flash before his eyes. His hands tremble as they clutch the edge of the desk. A low curse slips through his lips–an acknowledgement that he may be venturing into ideas that mock the boundaries of rational thought. Carter's upbringing in the disciplined halls of academia drilled into him the sanctity of linear logic, peer-reviewed validation, and skepticism toward anything remotely unquantifiable. His parents always prided themselves on their logical natures, molding him into a scientist wary of anything but the tangible. This new reality–one where consciousness itself could unravel the universe–feels more like an affront than a discovery. Still, he can't shake the intrigue. A memory of his late friend, a brilliant mind undone by his inability to reconcile his inner demons, flickers through his mind. Is he, too, teetering on the edge of madness?

As he reviews the swath of staggering data, an existential crisis gnaws at him. What if the universe isn't as ordered as he thought? What if Jaxon, in his chaotic brilliance, has tapped into something that fundamentally shifts human understanding? His palms grow clammy, and he dabs them against his lab coat, muttering disjointed thoughts about the undulating patterns. His mind oscillates between skepticism and belief, caught in a web spun by the equation.

Out in the hallway, the sterile white walls reflect the harshness of fluorescent lights. Carter spots a colleague, Dr. White, adjusting a lab coat, an emblem of rigorous scientific discipline. "Miles, you're still at it?" White's tone carries the familiarity of years of camaraderie, tinged with an underlying hint of concern.

Carter's touchstone for sanity had always been the unwavering adherence to the scientific method, a bastion against the irrational. White's words evoke a tether pulling him back to this sanctuary, yet he feels himself resisting. "It's complicated, Martin. Jaxon's work, it... it defies conventional metrics."

"Stick to the method, Miles," White urges. "Don't let this nonsense sidetrack you. We deal in facts, not abstractions."

That bite of reality digs deep. Carter nods absently, words failing him as White walks away, leaving him to his internal maelstrom. He returns to his lab, but it feels more confined than comforting now. The question from his colleague lingers, casting a pall over his thoughts as he ascends the narrow staircase leading to the rooftop. Each step echoes a drumbeat of hesitation and resolve.

Reaching the rooftop, Carter is greeted by a night sky ablaze with stars. The urban hum below is a mere whisper against the expansive silence of the cosmos above. He tips his head back, letting the cold bite of night air ground him, wrapping his arms around himself as if he could hold together the frayed edges of his reality. In the vastness above, he sees patterns, fragments of the equation dancing in the starlight, taunting his understanding.

The universe stands relentless before him, an open canvas of mysteries far beyond human grasp. Yet, here he is, a man of science, rooted in empirical evidence, grappling with the divine possibility Jaxon's work suggests. The weight of his revelations bows his shoulders, his breath a visible mist dissolving into the ether. He realizes that knowledge isn't just an accumulation of data but an intertwined dance of perception and mystery.

A soft sigh escapes his lips, the sound swallowed by the boundless sky. His mind, once tethered to certainties, now flits between wonder and dread, grappling with the same unknowns Jaxon might have stared into. Gazing at the myriad stars, he acknowledges the profound realization dawning within: the convergence of science and mysticism could be the most intricate puzzle of all, perhaps even a pathway to understanding consciousness itself.

With a sense of both humility and awe, Carter embraces the duality of his existence. He is but a fragment in a greater cosmic lattice, a node in an infinite web of consciousness. As he turns to leave, his heart carries the weight of knowledge and the lightness of infinite possibilities, setting the stage for what may come next in his exploration of Jaxon's enigmatic equation.

Dr. Miles Carter sits at his desk once more, his office now bathed in the sterile, artificial light that offers no comfort. The lab, usually a haven of empirical certainty, feels more like a maze of questions with no clear exit. The equation, the simulations, the seekers' fragmented tales–all whirl in his mind, refusing to settle. His once steadfast belief in logic and order has been shaken, the foundation of his scientific worldview cracked under the weight of the unknown.

He pulls out his notebook, the pages filled with meticulous notes, diagrams, and hastily scribbled thoughts from the past few days. Each page represents a battle between his ingrained skepticism and the tantalizing possibility of something greater. The data from his simulations lie before him, anomalous results that defy explanation and yet beckon him to dig deeper. The mathematical models, once his refuge, now appear as gateways to realms he had never dared consider.

Carter closes his eyes, massaging his temples as he tries to find clarity amidst the storm of thoughts. His mind drifts back to the café, to the haunted eyes of the seeker who had trusted him with their fractured experiences. The weight of responsibility presses down on him, a reminder that his pursuit of knowledge is not just a personal quest but one that could have profound implications for others.

His thoughts are interrupted by a soft knock on the door. Startled, Carter looks up to see Dr. White standing in the doorway, concern etched on his face. "Miles, you've been at this for days. You look like you haven't slept. Maybe it's time to take a step back."

Carter forces a smile, though it doesn't reach his eyes. "I'm close, Martin. I can feel it. Jaxon's equation... it's more than we've ever imagined."

Dr. White steps into the room, his expression a mix of worry and skepticism. "But at what cost, Miles? You're pushing yourself too hard. The data is fascinating, yes, but it's also dangerous. You need to take care of yourself."

Carter sighs, leaning back in his chair. "I know, but I can't stop now. There's something here, something that could change everything we know about consciousness, about reality itself. I can't walk away from that."

White looks at him, his eyes softening. "Just... promise me you'll be careful. I don't want to see you lose yourself in this. We need you here, grounded in the real world."

Carter nods, appreciating his colleague's concern, but knowing deep down that he's already crossed a threshold. The real world, as he once understood it, has expanded beyond the confines of traditional science. The equation has opened doors that can't be closed, and he's compelled to see where they lead.

After White leaves, Carter returns to his work with renewed determination. He knows the risks, the potential for losing himself in the process, but the lure of discovery is too strong to resist. He dives back into the data, his mind racing with possibilities, each new insight bringing him closer to the edge of understanding–and perhaps to the brink of madness.

The hours blur as Carter delves deeper into the equation, pushing the boundaries of his own knowledge and sanity. The lab's hum becomes a background drone, a constant reminder of the world he's stepping away from. He can feel the equation taking hold, its patterns weaving into his thoughts, reshaping his perception of reality. The distinction between the tangible and the abstract begins to fade, leaving him suspended in a liminal space where anything seems possible.

Finally, exhausted but exhilarated, Carter steps back from his work. The data on the screen is incomprehensible to anyone but him, a chaotic mix of numbers and symbols that pulse with an eerie, almost sentient energy. He stares at it, feeling a connection that transcends the scientific, a pull toward something greater, something that defies explanation.

As he closes his eyes, a vision takes shape–a fractal universe, infinite and ever-expanding, where consciousness and reality intertwine in a dance of creation and destruction. He sees himself within it, a tiny speck of awareness navigating the vast expanse, searching for answers that may never be found. The vision fades, leaving him breathless and shaken, yet more determined than ever.

Carter knows he's on the precipice of something monumental, a discovery that could alter the course of human understanding. But with that knowledge comes a profound sense of responsibility, a burden he's not sure he can carry alone. He needs allies, people who can help him navigate the treacherous path ahead.

He reaches for his phone, his fingers trembling as he types out a message to Jaxon. The words come slowly, each one weighed down by the gravity of what he's about to propose. When he's finished, he hesitates for a moment before hitting send, the message a beacon of hope and desperation sent out into the void.

Carter sets the phone down and stares at the equation on the screen, the lines and curves now familiar yet still unfathomable. He knows the road ahead will be fraught with challenges, but he's ready to face them, ready to confront the unknown with the same rigor and curiosity that has driven him his entire life.

As the night deepens, Carter sits alone in his lab, the glow of the monitors the only light in the darkness. The weight of the equation presses down on him, but within that pressure lies the potential for something extraordinary. He takes a deep breath, steeling himself for the journey ahead, knowing that whatever he finds, there will be no turning back.

The night stretches out before him, an endless expanse of possibilities and dangers. And within it, Dr. Miles Carter begins to understand that he is not just a seeker of knowledge, but a guardian of something far greater than himself.

Chapter 10

THE GEOMETRY OF GRACE

Jaxon stands amid a small circle of seekers in the bustling community center, his eyes glistening with a mix of passion and unease. The seekers, a motley group of varying ages, backgrounds, and experiences, lean in with a collective hunger for understanding. The hum of animated conversations around them creates an electric atmosphere, each word Jaxon utters acting like a spark igniting curiosity among his listeners.
"There's a fractal nature to reality," Jaxon explains, gesturing with his hands. "It's not linear, not confined... It's an endless pattern, repeating and expanding, interwoven with consciousness itself."
The seekers nod, captivated. Some scribble notes feverishly, while others simply absorb the words, their faces reflecting a variety of emotions ranging from enlightenment to confusion. The community center, a former warehouse turned haven for seekers of knowledge, is alive with the scent of incense and the muted sounds of distant chanting, creating an eclectic backdrop for this gathering of minds.
As Jaxon voices the intricate details of his fractal equation, he senses a duality within himself–pride for his creation, yet plagued by the shadow of self-doubt, the memories of his downfall as a chef

haunting the edges of his consciousness. He scans the room, each face a reminder of his journey; some appear skeptical, others awestruck, all united by a thirst for the mysteries he has unearthed.

At that moment, the atmosphere subtly shifts. Allegra steps into the room, her presence demanding attention. She moves with purpose, her eyes scanning the crowd like a magnetic force drawing all eyes towards her. Her attire, a cascade of flowing robes adorned with intricate patterns, complements her commanding aura. Without uttering a word, she captures the room's energy, transforming it into a collective focus on her entry.

"Your insights are truly groundbreaking, Jaxon," Allegra's voice cuts through the hushed murmurs as she approaches him. "You've tapped into something extraordinary."

Her words are wrapped in charisma, each syllable polished to perfection. The crowd tightens around her, their previous fixation on Jaxon momentarily diverted to Allegra's radiant allure. Jaxon's guarded stance becomes palpable, his previous ease dissolving as he feels the weight of her scrutiny, the familiarity of her demeanor stirring an uncomfortable recognition within him.

"You know," Allegra continues, her voice imbued with an almost hypnotic cadence, "there are ways to make your equation more accessible. To bring its transformative power to the masses who truly need it."

Jaxon narrows his eyes, a mix of intrigue and alarm brewing in his mind. He is acutely aware of the seekers' reactions, their admiration for Allegra evident in their gazes and awed whispers. Each subtle shift in their posture reflects the magnetic pull she exerts, their collective energy swaying like elvers drawn to a beacon.

The community center, once a neutral ground for open discourse, now feels like an arena of underlying tension. The seekers, despite their genuine curiosity, seem swayed by Allegra's promise of greater enlightenment. Jaxon's thoughts whirl, questioning her intentions, feeling the edges of her charm dangerously close to a coercive grip.

Allegra stands closer, a confident smile playing on her lips. "You have such a profound gift, Jaxon. It would be a shame if it remained limited to small circles. Imagine the potential, the enlightenment we could foster together."

Her proposition hangs in the air, the crowd's murmurs creating a backdrop of silent expectation. Jaxon's pulse quickens, the mix of allure and threat in Allegra's words awakening memories of his own reliance on persuasion–both in seduction and in his culinary exploits. Through the murmurs, the scent of sandalwood incense thickens, mirroring the enveloping tension.

"Yes, but the risks..." Jaxon trails off, his words cautious, each one weighed with a deep-seated wariness.

Allegra's eyes sparkle with an intensity that is both seductive and unsettling. "Risks are inherent in all pursuits of knowledge," she responds smoothly, "but with the right guide, those risks can be mitigated. We can ensure that only those worthy, those prepared, access this enlightenment."

The seekers shift around them, their faces a blend of hope and trepidation. An older man with a weathered face nods thoughtfully, while a young woman, eyes wide with wonder, leans in closer. They hang on Jaxon's every word, expectations swelling with each breath he hesitates to take. The space feels charged, every flicker of the dim overhead lights casting shadows that deepen the sense of gravity in the air.

Jaxon's internal conflict intensifies. He sees the allure Allegra offers –a broader audience, greater impact–but he cannot shake the memory of seekers who faltered, their minds unable to grasp the fractal's vastness, spiraling into madness. He remains contemplative, scanning the room, sensing both the urgency and hope of the attendees.

Allegra, ever observant, steps even closer. "Think of the lives we could improve, the souls we could guide. Isn't that worth exploring?"

Jaxon swallows hard, his mind a crucible of conflicting emotions. He feels the room's anticipation pressing in, the seekers' animated discussions only a peripheral murmur to his deepening focus on Allegra's proposition. His caution, once a protective barrier, now feels like a battleground where the stakes are nothing less than the essence of his discovery.

As the seekers continue to chatter animatedly, their excitement undiminished, Jaxon remains locked in contemplation, the dynamic between him and Allegra laying the groundwork for an inevitable conflict.

Jaxon leads Allegra to a quiet corner in the community center, a haven from the cacophony of excited seekers. The side room is dimly lit, with the ambient murmur of discussions fading into a soft hum. The dimness cloaks them, intensifying the intimacy of their conversation.

Allegra sits gracefully, her poise and presence commanding attention even in the subdued light. Jaxon lingers at the doorway for a moment, a heavy sense of caution weighing him down, before settling into a chair opposite her. The soft glow from a table lamp casts shadows across her face, highlighting her striking features.

"Jaxon, imagine a community where your equation could be truly understood and integrated," Allegra begins, her voice a low, seductive murmur. "These seekers, they are on the cusp of a transformation. With the right guidance, your work could unlock new realms for them. Realms of enlightenment."

She speaks with an almost poetic fluidity, her words weaving a vision that pulls at the edges of Jaxon's imagination. He watches her intently, feeling the pull of her charisma but equally aware of the dangerous seduction it holds. Memories of seekers faltering, their minds unraveling, surface unbidden.

"My equation is not just a tool, Allegra," Jaxon says, his voice tempered with both hope and resolve. "It's a reflection of the observer's state. It's not something to be used lightly or by those unprepared for its implications."

Allegra leans forward, her green eyes locking onto his, searching for a chink in his armor. "And that's where I come in. I've seen what unchecked knowledge can do, and I understand the stakes, Jaxon. Think of it not just as sharing, but as guiding. Together, we could ensure the seekers approach it with the proper reverence and understanding."

Her words resonate with the part of him that craves connection and redemption. He feels the echo of his past failures mingling with the present urge to believe in something greater. Yet, he cannot shake the disquiet that Allegra stirs within him. Her confidence reminds him of his own misguided arrogance from his culinary days, a time when he believed anything was possible if one dared to experiment.

Jaxon falls silent, contemplation weighing his features. He pictures the equation's intricate fractals, the way they unfurl like the petals of a rare and dangerous flower. The beauty and peril interlace, a complex dance of geometry and chaos.

Sensing his hesitation, Allegra redoubles her efforts, her voice smooth and unyielding. "Jaxon, I respect your caution. But consider the potential—we could foster a renaissance of understanding, a new era where knowledge doesn't break minds but enlightens them. The seekers need this. The world needs this."

The intensity of her words fills the room, creating a tangible pressure that closes in on Jaxon. He remembers the seekers who did not have the fortitude to bear the weight of the knowledge, their consciousnesses shattered into a thousand fragments. The memory of their suffering solidifies his wariness, tightening his grip on the secret he holds.

"I understand your vision, Allegra," he says quietly, his voice firm with newfound clarity. "But there's too much at risk. The equation isn't merely an intellectual puzzle; it's a spiritual accelerant. Those unprepared for its truth could be consumed by its depths."

Their gazes remain locked, a silent conversation passing between them. Allegra's lips part slightly as if to protest, but she catches herself, understanding that her words have found their limit.

In the silence that ensues, the shadows grow more pronounced, accentuating the gravity of their exchange. Jaxon's heart beats steadily, each thud a reminder of his role as the equation's guardian. He senses Allegra's defiance tempered by a reluctant respect, an acknowledgment that their paths, for now, must diverge.

As their private conversation ends, the intangible boundary between them solidifies. Allegra rises gracefully, her resolve untouched yet now shaded with an awareness of Jaxon's steadfastness. Jaxon watches her leave, the weight of his decision settling heavily in his chest, knowing that the struggle between caution and enlightenment has only just begun.

Allegra's mind drifts back to that fateful night, the one where her path to boundless knowledge began. She is young, vibrant, and full of eager anticipation as she steps into the dimly lit room of the secretive esoteric seminar. The air is thick with the scent of swirling incense that weaves a tapestry of sandalwood and myrrh, a sensory guide into the realm of the unknown. Each corner of the room is adorned with eclectic decor–intricately woven tapestries, statues of deities lost to time, and candles flickering their enigmatic rhythms.

She glides through the crowd, her eyes taking in the diverse assembly. The participants are a mix of fervent seekers, skeptics, and self-proclaimed prophets, their faces illuminated by both hope and apprehension. As they gather around the speaker, their voices blend into a rising murmur of expectation and doubt.

The seminar leader, a mysterious figure cloaked in robes embroidered with celestial patterns, steps forward. They begin to unravel the secrets of a fractal concept similar to what Allegra would later encounter through Jaxon's work. The leader's words interlace with the room's subdued lighting, painting an image of reality as an infinite cascade of repeating patterns–each iteration a microcosm, each divergence a potential for enlightenment or madness.

Participants listen intently, their reactions as varied as the fractals themselves. Some are visibly enlightened, faces glowing with newfound understanding; they nod in unison, their minds seemingly merging with the cosmic patterns described. Others exhibit signs of distress, confusion settling into their features like shadow on parchment. Allegra's heart races as she witnesses a young woman clutch her head, eyes wide with the terror of comprehension turned into chaos. Another man collapses, body wracked by spasms as his consciousness grapples with the concept's weight.

Allegra observes keenly, absorbing the duality of enlightenment and peril that esoteric knowledge can bestow. She watches as minds fracture and fragment, some descending into madness as the universe's fractal geometry infects their psyche. Yet, amid the chaos, she also sees glimpses of divine connection–people transformed, their spirits seemingly soaring across dimensions tethered by sacred geometry.

She stands in the darkened room, the air vibrating with unmanifested energy, and realizes the monumental responsibility that comes with such knowledge. The symbols and scars of this revelation etch themselves into her memory, defining her future resolve to guide others. Was

The flashback jerks to an end as Allegra steadies herself, her present consciousness snapping back into clarity. The dim light of the community center bathes her in a sepia glow, reflections of her younger self mingling with the person she has become. She understands now more than ever the delicate balance required to navigate between illumination and insanity.

"I can guide them," she asserts, her voice firm yet laden with the weight of responsibility. The memory reinforces her determination and solidifies her belief in her ability to harness the equation's potential for the greater good.

The participants in the room continue to exchange eager conversations, unaware of the tempestuous history and profound realizations that fuel Allegra's current conviction. As she lifts her head, her resolve reflects in her eyes, shining with a mixture of wisdom and ambition. The scent of anticipation and the low hum of whispered theories envelop the room, amplifying the stakes that lie ahead.

Allegra stands with a sense of purpose, the duality of her past and present converging. She embodies the promise of enlightenment and the shadow of chaos, and she is resolute in her path to navigate both with unwavering certainty. The room's ambient murmurs echo her internal vow–to wield esoteric knowledge responsibly and to guide others through the labyrinthine journey of self-discovery.

Allegra leans closer, the subtle fragrance of incense and sandalwood lingering in the air as her voice drops to a whisper, "Jaxon, think of the mysteries still locked within your equation. You've only scratched the surface. Together, we can unlock so much more."

Jaxon's mind races. The allure of deeper knowledge tugs at him, a tantalizing whisper against the backdrop of his caution. He sees Allegra's green eyes shimmer with persuasive intensity, the promise of enlightenment gleaming in them. The community center hums softly in the background, seekers dispersing, their animated discussions a distant drone.

As Allegra's words sink in, Jaxon feels the familiar tug-of-war between temptation and responsibility. He thinks of the seekers who faltered, their minds unable to grasp the enormity of the equation. Their fragmented consciousnesses haunt him, lingering like phantoms in his thoughts. The pain caused by unwary exploration is a stinging reminder of the dangers inherent in his work.

Allegra presses on, sensing his wavering, her voice a melodic current. "The equation holds immense potential. Imagine the transformation we could bring to the world. Enlightenment, accessible to all."

Jaxon hesitates, his brow furrowing. Memories of seekers lost in the labyrinth of realities flood his mind. The fractal equation is more than a key; it's a double-edged sword. He feels the weight of their fractured souls, the chaotic resonance of their failed journeys echoing in his ears. The ambient noise fades as his resolve crystallizes, the gravity of his mission settling heavily within him.

Drawing a deep breath, Jaxon's voice trembles with the burden he has carried. "Allegra, the equation isn't something that can be shared lightly. It's not just knowledge; it's a reflection of the state of the observer. People aren't ready. I can't... we can't risk it."

Allegra's eyes flash with a mix of frustration and understanding. She takes a step back, processing his words, her charismatic façade momentarily slipping. The tension between them crackles, electric and tangible, as the crowd continues to drift away, leaving pockets of silence in their wake.

The community center's dim lighting casts elongated shadows, amplifying the weight of Jaxon's decision. Allegra's persuasive insistence lingers in the air, a compelling force that challenges his newfound clarity. But Jaxon stands firm, his eyes holding a steely determination born from hard-earned wisdom.

The last of the seekers depart, their footsteps echoing softly on the polished floor. Allegra's expression hardens, a mix of disappointment and grudging respect. She nods slowly, acknowledging the unspoken finality of Jaxon's resolve. Their paths, once converging with promise, now stand at a crossroads marked by his choice to protect the essence of his work.

Jaxon watches as Allegra turns away, her figure blending into the dim light. The gathering disperses, the atmosphere heavy with the lingering tension of what might have been, the air thick with unsaid words and unresolved conflicts. Jaxon's heart pounds with the weight of his decision, the gravity of safeguarding his creation from being a mere tool of allure and ambition.

Now, standing alone amidst the fading murmurs of seekers, Jaxon feels a strange sense of peace mingled with the tight uncertainty of what's to come. This moment—a blend of triumph and foreboding—marks a pivotal turn in his journey, a beacon and a warning for future encounters. The silence that follows is profound, heavier than the air in the room, promising a clash yet to unfold, a struggle for control over the infinite depths of his equation.

Chapter 11

CIRCLES OF LIGHT, CIRCLES OF DARK

Jaxon opens the door to a gaunt figure, eyes wide with both fear and fervent curiosity. The young mathematician stumbles into the room, clutching loose papers filled with scribbled equations that flutter like trapped birds. His voice trembles as he speaks, words tripping over each other in a desperate attempt to convey his thoughts.

Adjusting his worn leather jacket, Jaxon takes a step back, the mathematician's energy crackling between them. The living room, usually a sanctuary of solitude, now feels charged with an electric tension. The faded wallpaper and old furniture seem to close in, reflecting Jaxon's mounting anxiety.

The mathematician paces, gesturing wildly at his notes, his voice rising and falling with each revelation. Jaxon nods absently, attempting to grasp the flood of information. Every phrase, every symbol on the fluttering pages, feels like a demand, an accusation against the fragile stability Jaxon clings to. The air grows heavy, each breath harder to take, as the chaos of the young man's thoughts penetrates deeper into Jaxon's psyche.

Jaxon's mind flashes back to the calm precision of his kitchen, where each ingredient had its place, each movement was deliberate. There, amidst the symphony of sizzling pans and aromatic herbs, he found peace. Here, in this maelstrom of human desperation, peace seems but a distant memory.

The front door creaks open again, and suddenly, the room is filled with more seekers, each one more desperate than the last. Their faces blur into a cacophony of need, voices overlapping in frantic pleas for understanding. They crowd closer, their bodies pressing into Jaxon's once personal space, transforming his sanctuary into a suffocating hive of longing and confusion. The weight of their combined energies feels like a physical force, pressing down on his chest, threatening to crush him under its enormity.

Images of his past life flicker through his mind—flambé dishes, the controlled chaos of a dinner rush, the thrill of creating culinary art that left patrons in awe. The tangible, immediate satisfaction of his work now juxtaposes sharply with the abstract, infinite complexities of the knowledge he's unlocked. The contrast is jarring, and he wonders if he's truly ready to guide these souls when he himself is still fumbling for answers.

Feeling the walls closing in, Jaxon retreats, pushing his way through the mass of bodies. Every touch, every brush of a hand feels like another weight added to his burden. He stumbles into a quieter part of his home, a room far removed from the din outside. Here, the air feels lighter, the silence almost sacred. He closes the door behind him, leaning against it for support.

The noise from the seekers becomes a muted hum, like a distant, angry sea. Jaxon closes his eyes, taking a deep breath, savoring the fleeting moment of solitude. Each inhale and exhale is an anchor, pulling him back from the brink of panic. The muffled sounds of their voices outside remind him of the immense responsibility he now bears.

The weight of the seekers' expectations settles heavily on his shoulders. Can he truly guide them, or is he leading them into an abyss? Doubts gnaw at him, memories of his darkest days resurfacing—the alcohol, the drugs, the nights lost in a haze trying to escape his own mind. He fears that the insights he's gained might not be enough, that he might lead these seekers down a path of no return.

In this moment of calm, he clings to the image of his former self, a chef who once found solace in the simplicity of his craft. He realizes that while his culinary creations once brought fleeting joy, the knowledge he now possesses has the potential to transform lives fundamentally. But with that potential comes great risk.

Jaxon takes another deep breath, feeling the weight of his decisions bearing down, understanding that while he can shine a light on the path, each seeker must walk it themselves, navigating their own inner struggles. For now, he seeks only a moment of peace, hoping to gather the strength to face the torrent of human desperation waiting just beyond the door.

Jaxon settles himself on the floor of the secluded room, his eyes closing as he tries to block out the cacophony coming from the seekers outside. He focuses on his breath, each inhale and exhale a fragile attempt at reclaiming serenity. The room, bathed in soft, golden light filtering through the single window, feels like a fragile sanctuary amid the storm encroaching upon his life.

He hopes that solitude will offer him the clarity he's desperately searching for. The house, once a fortress of his thoughts, now feels like a shrinking shell under the persistent invasion of desperate minds. Jaxon's equation has become both a beacon and a curse, drawing those who yearn for answers, yet threaten to drown him in their collective chaos.

Muffled voices from the next room seep into his sanctuary. The seekers, driven by fear, desperation, and relentless curiosity, assemble like a storm battering at the walls of his mind. Phrases drift in: "the key to everything," "revelation," "enlightenment." Jaxon's breath hitches as the weight of their expectations bears down on him, each word an emotional tidal wave crashing against his fragile peace.

His mind, a once-clear conduit of cosmic patterns, feels cluttered by the intrusion of so many foreign desires. The societal hunger for knowledge and answers has morphed into a frantic scramble, tearing at the fabric of reality and sanity. Jaxon senses their needs, their fears—each like a jagged shard of glass tearing through his newfound understanding. The world outside his door teeters on the edge between enlightenment and madness, each seeker a reflection of the precarious balance.

Jaxon tries to sink deeper into his meditation, but the intensity of the emotions surrounding him disrupts his focus. Moments of clarity slip through his grasp, fleeting as the autumn leaves outside his window. His heart begins to race, a drumbeat of unease amplifying with every passing second. He feels the longing of the mathematician who was the first to arrive, the man's mind a tangle of numbers and existential dread.

A wave of frustration wells up within him, echoing the cacophony outside. Jaxon wonders if he has become another drug to these seekers, each one desperate for a hit of enlightenment that might forever alter their lives yet might also shatter their sanity. He fears the echoes of his past, the addiction that once enslaved him, rearing its ugly head again but this time through others' dependency on his discoveries. The ethical weight of sharing his equation presses harder against his chest.

Could he be leading them to their own destruction? The thought claws at his conscience, making his breath hitch. He considers Maya and the unique blend of companionship and rivalry she brings into his chaotic existence. Her presence intensifies his internal conflicts, making him yearn for an impossible reconciliation between solitude and connection. The connection he has with Maya is complex, danced around the edges of camaraderie and competition.

Jaxon opens his eyes, the soft focus of the room doing little to ground him. The seekers' emotions batter against his consciousness like a thousand tiny fists, each demanding entry. He can no longer deny their effect; their turmoil has become his own. In destroying each physical equation, he intended to safeguard against misuse, but he had not foreseen the spiritual wounds this awakening would tear open.

A deep breath in, he forces himself to look around the room, trying to anchor himself in the present. The worn tapestry on the wall, the quiet hum of his electric heater, even the slightly cracked windowpane–all these become lifelines pulling him back from the precipice of overwhelming anxiety. His repose must be maintained, but he knows that the journey ahead involves navigating these turbulent waters without abandoning his essence or the truths he aims to share.

In this liminal space of silence and noise, Jaxon grapples with the duality of his existence. He is both key and gatekeeper, a celestial waypoint for human consciousness and yet a soul fraught with fears and fragmented pasts. The knowledge he carries is a fire—capable of illuminating the darkest corners of human understanding, but just as prone to burn those unprepared to handle its searing truths. Reclaiming his composure, he resolves to uphold the sanctity of his discoveries and guide those seekers not through dictation, but through a shared journey of enlightenment.

Even here, amid the surging tide of seekers' voices, Jaxon finds a burgeoning sense of clarity and determination. He closes his eyes once more, inhaling deeply, readying himself to face the storm that brews just beyond the door.

A frantic seeker barges into the living room, his eyes wild with desperation. Jaxon barely has time to step back before the man's hands are gripping his shoulders.

"You have to tell me, Jaxon!" the seeker demands, voice cracking with urgency. "I need to understand the equation. It's the key, isn't it? It's the answer to everything!"

Jaxon attempts to extricate himself gently from the man's grip, feeling the raw energy radiating off him. The living room is a tight, cluttered space–bookshelves sag under the weight of volumes on chaos theory, quantum mechanics, and sacred geometry. Various diagrams and sketches plaster the walls, half-hidden beneath layers of post-it notes scribbled with fractal patterns and mathematical symbols, a testament to Jaxon's intense and fervent journey.

"Please, calm down," Jaxon says, guiding the seeker toward a chair. The slight tremble in his hands betrays his own fraying nerves. "Understanding the equation takes time. It's not something you can rush."

"But I need it now!" the seeker's voice escalates, his eyes flashing with fear and frustration. "I've lost everything–my job, my family. This is my last hope. You can't keep it from us!"

Jaxon's mind races. He can sense the man's desperation, a mirror of his past battles with addiction, where his need for a fix overshadowed all reason. This man's intense craving for instant enlightenment evokes a startling parallel. Jaxon understands all too well the path of destruction one can pave in pursuit of solace.

"I'm trying to help you," Jaxon insists, his tone growing firmer. "The equation isn't just a solution; it's a gateway. If you approach it without the right mindset, it can shatter you, not enlighten you. You have to prepare yourself mentally and spiritually."

The seeker's agitation only grows. He stands up, knocking over a stack of books that crash to the floor with a resonant thud. "Don't patronize me, Jaxon! You sit here, hoarding knowledge while the rest of us suffer. Who are you to decide who gets to be saved?"

Jaxon's patience snaps. His heart pounds against his ribs, the room around him seeming to grow smaller, closing in on him with each passing second. Memories of his fall from grace flood his mind–a once-renowned chef now grappling with cosmic equations in a desperate bid to make sense of his fragmented reality.

"I'm no one's savior," Jaxon replies, his voice cold and resolute. "But I know what happens when you dive into something you're not ready for. The equation is a reflection of you. If you approach it broken, it will only break you further."

The seeker's face contorts with bitter resentment. "You think you're better than us? You think your past makes you some sort of gatekeeper to the divine?"

Jaxon raises a hand, exhaling deeply to steady his tumultuous emotions. He struggles to shake off the ghosts of his past, the moments when he hesitated and faltered, the years lost to numbing his pain with substances. The weight of his responsibility bears down on him like an anchor, a constant reminder of the fine line he walks between wisdom and madness.

"I'm not better," Jaxon says slowly, "I'm just as flawed as anyone. But I've seen what happens when that equation overwhelms an unprepared mind. It's not about withholding knowledge–it's about protecting you from yourself."

The seeker, now visibly trembling, locks eyes with Jaxon, hatred mingling with desperation. "You'll regret this," he mutters through clenched teeth, storming out of the room and slamming the door behind him.

As the echoes of the door reverberate through the silent house, Jaxon sinks into a chair, his body feeling as though it has been wrung dry of all strength. The room is a chaotic mess of overturned books and scattered papers, mirroring the turbulent storm within him. It's a battle he faces repeatedly–balancing the need to guide others with the imperative to protect them from the seductive dangers of half-understood wisdom.

He looks around, the shadows of his earlier self lingering in the corners. The culinary studio now serves as a metaphor for his life– a space where raw ingredients of experience and knowledge are transformed into something meaningful. It's his sanctuary, yet even here, the outside world creeps in, demanding more than he feels able to give.

Jaxon closes his eyes, exhaling slowly, the weight of the recent confrontation settling heavily on his shoulders. He knows that this is merely the beginning of the trials to come. The seekers won't stop–each one brings their own turmoil to his door, seeking salvation in his equations, often with reckless abandon.

Gathering his resolve, Jaxon acknowledges that tonight's experience only affirms the need for a measured approach. His journey towards understanding and responsible knowledge-sharing must remain steadfast, despite the chaos it incites. He can only hope that, with time, his seekers will learn the value of the path through self-discovery. For now, Jaxon must continue to navigate the delicate balance of guiding others while remaining true to the enlightenment he has painfully wrought from his tumultuous existence.

After the confrontation, Jaxon sits alone in his study. The room is a refuge amidst the maelstrom that his home has become. Heavy curtains block out the waning light of day, casting shadows that dance across the wooden floor. The scent of pine from the bookshelves mingles with the faint aroma of old leather and ink, grounding him in familiar comforts.

He lowers himself onto a plush chair, the leather creaking under his weight, and opens his journal–a sanctuary for his thoughts. The blank page beckons, a canvas for the turmoil swirling inside him. Picking up his pen, he begins to write, each word a tentative step into the labyrinth of his mind.

"What does it mean to carry such knowledge?" he wonders silently, the pen gliding across the paper. The memories of his past life, filled with the vibrant clamor of kitchens and the accolades of being an avant-garde chef, now seem like ghostly echoes. His fall into addiction wasn't just a tumble into substance abuse–it was a freefall into oblivion where his genius had once thrived. Being struck by lightning saved him physically but unearthed a new torment–carrying insights that could unravel minds.

The seekers' desperation floods back into his thoughts, each face a mosaic of fear and longing. He writes, "Knowledge isn't gold–it's weight, a burden." This equation he has unveiled to the world isn't merely numbers; it's a doorway to the infinite. Yet, without the wisdom to wield it, that infinity can consume the unwary. His mind tugs him back to the raw moments of his addiction, how it fractured his relationship with Angelica. Her face, a mix of hope and disappointment, haunts him. How many times had she tried to save him? And now, how could he save these seekers from themselves?

Jaxon's thoughts drift to Maya. She's a paradox–both a muse and a challenge. Her passion and philosophical curiosity push him towards greater understanding, yet her drive often borders on aggression and rivalry. Their conversations ring in his ears, each debate a dance between revelation and frustration. She is a mirror reflecting his own indecision and fear back at him, her quest for enlightenment entwined with his. He scribbles down, "Maya challenges my clarity, but also propels me forward. I need to understand her to understand myself."

The journal becomes a rhapsody of distress and resolution. Jaxon records the ethical quake that has been shaking his core. Sharing this equation could enlighten or destroy–something he now understands all too well. The seekers, clamoring at his door, each carry their own fragment of chaos, and in this equation, that chaos finds fertile ground to grow. "This knowledge is not a gift to be given lightly," he notes, feeling the weight of his words.

His thoughts shift to the concept of enlightenment. True wisdom isn't in the superficial acquisition of profound knowledge but in the arduous journey of self-realization. He remembers his culinary mentor who taught him that a perfect dish wasn't in the ingredients alone but in the hands that balanced them, in the heart that understood their essence. Enlightenment, he realizes, mirrors this balance–each person must find it within, not expect it to be delivered like a recipe.

Jaxon's pen pauses as the gravity of his role sinks in. Can he guide these seekers? The frantic man from earlier flashes in his memory, his aggressive desperation a stark warning of the perils that lie ahead. Their clash was a manifestation of more than just knowledge; it was about readiness, about the seeker's inner world.

He writes more feverishly now, the pen struggling to keep up with the torrent of ideas. "I must become a guardian. Not of the knowledge itself but of the journey towards it." Each word crystallizes his new path. He sees now that his duty isn't to dole out answers but to foster questions, to nurture the seekers' own quests for understanding.

The journal captures his transformation from a man overwhelmed by external pressures to one who sees the path before him with a stark clarity. He reflects on Angelica again, recognizing that to mend their fractured bond, he must first find harmony within himself. His sister had always believed in him, and now, equipped with this insight, he feels an urgency to reach out to her, to bridge the chasm his addiction had dug.

With a final flourish, Jaxon closes the journal. The room, once a confining space, now feels expansive, filled with the echoes of his realizations. He exhales deeply, the act simple yet profound, as if breathing in the essence of his newfound resolve.

He stands, the journal held close to his chest, and moves towards the door. The murmur of seekers outside remains, but within him, a seed of purpose begins to grow. Enlightenment, he now understands, is not a treasure to be distributed but a journey to be undertaken. And he is ready to guide those who truly seek, those prepared to confront their own inner chaos.

Jaxon closes his journal, feeling a newfound sense of clarity and determination.

Chapter 12

THE FLAVOR OF MEMORY

Jaxon steps into the underground meeting space, a dimly treeith anticipation, charged with the scent of paint and the ozone tang of electricity, intermingled with the musky aroma of bodies huddled close in intellectual fervor. Conversations overlap, fragmented and intense, punctuated by bursts of heated debate. The space is a strange amalgamation of the esoteric and the empirical—where art and scientific instruments coexist in a chaotic harmony, creating a vibrant yet disconcerting atmosphere.

To Jaxon, these gatherings are both a refuge and a crucible, a space where the esoteric and the empirical collide. The underground culture that has sprung up around his equation reflects a broader shift toward holistic exploration, where intuition and logic are given equal weight. Recently, his fame has drawn a spectrum of individuals—scientists disillusioned by the limitations of their fields, mystics seeking transcendence, and curious minds yearning for a glimpse beyond the veil. Their shared pursuit of understanding creates a unique culture, a microcosm reflecting a broader shift towards holistic exploration where intuition and logic are given equal weight.

As he moves through the room, a loud argument catches his attention. A seeker, animated and intense, stands in the center, gesticulating wildly as they debate the implications of Jaxon's work. The atmosphere is electric, tendrils of intellectual energy crackling in the air. The seeker's voice rises above the din, "You're not seeing the full picture! It's not just mathematics; it's a map of consciousness!"

Jaxon feels a knot tighten in his stomach, the familiar twinge of anxiety mingling with a deep-seated responsibility. He's proud of the equation's beauty but dreads the consequences of its misuse. The idea that his work could unravel minds weighs heavily on him.

Approaching the edge of the room, he finds a seeker whose presence offers a degree of comfort. This individual, known for their balanced perspective, often serves as a grounding force amidst the chaos. Jaxon leans in, his voice calm yet betraying his concern, "What do you think of all this? Are they pushing too far too quickly?"

The seeker meets his gaze, eyes reflecting both wisdom and the weight of their mutual understanding. "There's a fine line between enlightenment and madness, Jaxon. Your equation is powerful, but not everyone is ready for what it reveals."

Their words echo in Jaxon's mind, weaving through his thoughts like a warning. His fame has brought this underground culture to the brink of something monumental, but also dangerously uncharted. He turns his attention to the far side of the room where two seekers, emboldened by their perceived understanding, prepare to delve deeper into the equation.

Their confidence radiates, infectious yet unnerving. Jaxon watches, a sense of foreboding creeping up his spine. The room shifts, a collective breath held as the seekers begin their advanced exploration. They channel their energies into deciphering the equation, fingers tracing patterns in the air, minds reaching into dimensions that Jaxon himself can barely grasp.

The atmosphere grows tense, an undercurrent of anxiety threading through the room. Whispers ripple among the crowd, fragmented snippets of concern and curiosity, "Are they ready for this?" "What if they get lost in it?" "Can Jaxon control the chaos?"

Jaxon's heart races, every muscle tensed as he assesses the situation. He senses the seekers' consciousness teetering on the edge, a precarious balance between revelation and ruin. The room feels claustrophobic, the walls closing in as the equation's complexity exerts its pressure. Each pattern they draw seems to fold into itself, a fractal bloom of possibilities that could unravel minds unprepared for its intensity.

He feels the collective gaze of the room upon him, a silent plea for guidance amidst the uncertainty. The seekers' hubris is a mirror of his own past mistakes, their ambition both admirable and terrifying. He steps forward, voice steady but laced with urgency, "Take it slow. You're venturing into dangerous territory."

The seekers pause, glancing at Jaxon with a mixture of respect and defiance. "We understand the risks," one of them asserts, their hands trembling ever so slightly. "We need to know what lies beyond." The room holds its breath.

Jaxon remains poised on the knife's edge of action, caught between his duty to protect and the unrelenting pull of curiosity that drives them all. He senses the moment teetering, the potential for enlightenment shadowed by the specter of dissolution.

The underground meeting space vibrates with a palpable energy, the dim lighting casting exaggerated shadows across the walls. The air is thick with the scent of incense and the murmur of intellectual debates. Seated in small clusters, the seekers seem to be held together by a fragile thread of curiosity and desperation. Their conversations, a mélange of fragmented statements and gesticulations, are punctuated by moments of tension. One seeker, face flushed and eyes wild, clutches a piece of paper filled with fractal equations, waving it dramatically as he bellows out his interpretation.

Jaxon stands at the periphery, a silent observer to the chaotic scene. The room hums with a frenetic energy that gnaws at his sense of peace. His eyes scan the seekers, noting the tension that grips them as they delve into the complexities of his equation. He catches sight of one seeker on the edge of the room, separate from the furore, their face creased with confusion and fear. Jaxon steps forward, the nuances of light and shadow playing across his features, as he approaches in hopes of extending a comforting presence.

The atmosphere shifts dramatically when two seekers, evidently more confident than their understanding justifies, decide to embark on an advanced exploration of the equation. They gather around a space filled with scattered papers and electronic tablets, their voices rising in excitement and growing louder with each exchanged word. The crowd gravitates toward them, a mix of enthusiasm and skepticism evident on their faces.

The seeker at the edge, already teetering on the brink of emotional stability, suddenly lets out a sharp cry. With a gasp, they collapse to the floor, hands clutching their head as if trying to contain an internal storm. The room bursts into a cacophony of panic, seekers rushing to help but their efforts only adding to the confusion. Shouts and hurried movements echo against the walls, amplifying the sense of claustrophobia.

Jaxon pushes through the throng, his heart pounding in his chest. He kneels beside the fallen seeker, trying to remain a pillar of calm amidst the chaos. His voice, though firm, is soothing as he tries to reassure them, but the seeker's erratic breaths and the wild flicker in their eyes suggest a profound disconnect from reality.

"It's going to be alright. Focus on my voice," Jaxon says, but his own words sound hollow in the din of panic.

The seeker's condition deteriorates rapidly, their incoherent shouts slicing through the frantic babble of the crowd. "I see... worlds... folding, collapsing into each other!" they scream, eyes wide with terror. The raw fear in their voice sends chills through everyone present. A frenzied spell ripples through the room, causing the crowd to withdraw in fear.

Jaxon's mind races, wrestling with an overwhelming sense of dread and responsibility. He feels the weight of each seeker's gaze and the oppressive confines of the room pressing in on him. The walls seem to close in, the once expansive space now claustrophobic, suffocating. The seekers' movements grow sluggish, their inner turmoil mirrored in the dim lighting that casts deep shadows, making the room feel smaller, more fragmented.

Among the crowd, Maya's face reflects an internal battle–a mixture of envy and desperate support for Jaxon's leadership, her eyes filled with a struggle that words could not articulate. Dr. Miles Carter's skepticism sharpens, his analytical mind challenged by the overwhelming evidence of the surreal unfolding around him. He stands still, his normally composed facade cracked, uncertain of what to make of the chaos. Allegra Frost's authoritative demeanor falters, her usual control slipping as she grapples with the magnitude of what the seeker is experiencing.

As the seeker's agonized cries grow increasingly incomprehensible, the room vibrates with tension. Shadows flicker on the walls, ghosts of the seekers' inner turmoil. Relationships strain under the pressure–initial camaraderie giving way to confusion and terror. Trust erodes, suspicions bubbling beneath the surface.

Jaxon, feeling the trust and resentment mingling in the air, holds onto a fragile thread of hope. But the weight of guilt presses down on him with crushing force, each agonized scream a reminder of the burden his revelations have placed on fragile minds. Seekers begin withdrawing, their movements slow and apprehensive, each step a retreat from the fracture lines spreading through their collective psyche.

The emotional chasm that opens in the wake of the seeker's breakdown is deep, echoing with the sound of shattered trust and broken dreams. Seekers who previously thrived on shared curiosity now grapple with fear–their relationships strained, their pursuit of knowledge tainted by the haunting specter of potential madness.

The room's anxiety peaks, turning the atmosphere oppressive. Jaxon, the center of the storm, feels the fulcrum of the seekers' shifting perceptions, realizing the profound impact of his creation. The collective fear crescendos as the seeker's condition worsens, the air heavy with unspoken dread and mutual suspicion.

As tension coils around him, Jaxon senses the claustrophobia that envelops the room, choking out the last vestiges of hope.

Jaxon stands rooted to the pavement, a statue of shock and disbelief amidst the rising chaos. Neon lights, flickering from a nearby convenience store, cast eerie shadows across the faces of the seekers huddled on the dimly lit street. The cacophony of panic reverberates, clashing with the distant wail of sirens that herald the arrival of medical personnel. The scene before him is a dizzying blur of motion–bodies darting, voices raised, hands trembling.

The sharp smell of antiseptic mingles with the more pungent, acrid odors wafting from the underground meeting space. Seekers whisper in hushed, urgent tones, their words heavy with dread. "What have we unleashed?" one voice mutters, frantic eyes darting toward Jaxon. "The equation... it's too dangerous," another insists, voice quivering with fear.

Jaxon's mind drifts and stumbles through fragmented memories of his own past. The sickening lurch of withdrawal, the hollow ache of loss, the relentless grip of addiction that once held his soul in a vise. His memories crash like waves: the shattered glass from a thrown bottle, the haunting emptiness in Angelica's eyes as she turned away from him, unable to witness his spiraling descent. Now, here he stands, the architect of a new kind of chaos, one that fractures minds and disrupts realities.

"Someone do something!" a seeker pleads, their voice cracking the fragile night air. A stretcher emerges from the ambulance, its wheels clattering on the uneven street, a stark reminder of Jaxon's culpability. He feels each breath heavy in his chest, the weight of guilt pressing down like an unseen hand. His equation, once a beacon of intellectual brilliance, now seems a dark harbinger of madness.

Trying to wrest control over his spiraling emotions, Jaxon's inner monologue races. How could I be so irresponsible? The fractal patterns, so alluring and beautiful in theory, have become labyrinths of endless torment. His thoughts mirror the spirals in the equation, looping endlessly back to his monumental error.

He can still feel the chill of the knife's edge that left a scar on his cheek, a makeshift punishment from his culinary past now echoed in a greater cosmic retribution. The seekers' faces are a gallery of horror and confusion, each expression etching guilt deeper into his consciousness. This is not just a wound on one individual but a collective fissure in a group that sought enlightenment and instead found despair.

Emergency responders lift the seeker, their careful motions cloaked in the eerie efficiency of well-trained hands. The streetlights dance across the pale, sweat-drenched face, now contorted in a silent scream. Jaxon's heart clenches, feeling each fragment of the seeker's terror as if it were his own. His equation, that delicate synthesis of geometry, chaos, and quantum entanglement, had become an instrument of profound suffering.

A fleeting thought of Angelica breaches his consciousness, her face painted with the hues of concern and love he has rarely seen since his descent into addiction. There is a pang of grief as he acknowledges her absence now, a chasm of emotional isolation that seems to deepen with each passing moment. His sister, whose compassionate eyes once held him steady, is now a ghost in his life, a presence felt but unseen, her absence a stark reminder of his fractured existence.

As the stretcher is loaded into the ambulance, Jaxon turns inward, confronting the reality he has wrought. He sees the seekers, each one a reflection of his own inner turmoil, a mirror to his past that he can no longer ignore. Their questioning eyes challenge him to explain, to justify, but words fail him. The guilt is a tangible weight, a dark, cloying presence that seeps into every thought, every breath.

The ambulance door slams shut, a finality that echoes through the night. Jaxon's surroundings blur, the world shrinking to the narrow alleyway where he now stands alone. He grapples with his thoughts, the undeniable truth that he must face the ramifications of his creation. He has unleashed something far beyond his control, a Pandora's box of cognitive dissonance that leaves him with a profound sense of responsibility.

There is a moment of clarity amidst the chaos, a sliver of understanding that pierces through the fog of his mind. The equation is not merely a tool; it is a reflection of the observer's state, amplifying the internal chaos into the external realm. His eyes lift to the dim glow of streetlights, their luminescence a fragile reminder of hope amidst despair. He cannot undo the past, but he can confront it, engage with it, and seek a path forward that mitigates the harm his work has caused.

Jaxon's shoulders slump as he watches the ambulance speed away into the night, the red and blue lights fading into distant pinpricks. The street around him feels cold, the voices of the seekers reduced to a murmur. He is left in the void of his thoughts, the silence around him a stark counterpoint to the clamor within. The weight of isolation presses hard upon him, yet within that solitude, a seed of resolution begins to germinate.

His mind is a swirl of self-reflection, and as he stands there, he begins to acknowledge the path he must tread. The road to redemption is fraught with challenges, but it is a journey he must undertake. As the last echoes of the ambulance siren fade, Jaxon takes a deep breath, letting the night air fill his lungs. The journey toward understanding and reconciliation has only just begun.

Alone in his dimly lit apartment, Jaxon paces restlessly, each footfall a muffled thud against the worn wooden floors. The room, sparsely furnished, bears echoes of his culinary past–copper pans hanging like lost relics, a row of empty spice jars lining the countertop, their once-vibrant scents now ghosts in the air. Outside, the distant hum of the city makes him acutely aware of his isolation, amplifying the heavy silence within.

Images of the seeker's collapse flood his mind: the stark terror in their eyes, the contorted body gasping for air, the room's frantic energy as others tried and failed to offer help. These visions replay in a relentless loop, each one a stab of guilt and helplessness. His heart races, beads of sweat forming on his brow. He sinks into a chair, feeling the weight of responsibility crushing his chest.

He grabs a worn notebook from the table, its pages already filled with chaotic equations and fragmented thoughts. His hand shakes as he scrawls furiously, the ink bleeding into the paper. "I've unleashed something beyond understanding," he writes, the words a desperate attempt to grasp the enormity of his actions. Each sentence is a fragment of his fear and guilt, a documentary of his inner turmoil.

His thoughts return to Angelica, his estranged sister. How many times had she tried to pull him out of the darkness of addiction? Her voice echoes in his mind, gentle but firm, pleading for him to see the potential within himself. Every memory of her is tinged with regret and longing–a reminder of the bond they once had and the pain he caused her. He remembers her eyes, filled with a mix of love and sorrow, and the guilt gnaws at him like a ravenous beast.

The notebook slips from his grasp, landing with a soft thud. He leans back, closing his eyes, and takes a deep breath, attempting to calm the storm inside. The air around him feels charged, as if the universe itself is demanding his attention. He decides to meditate, hoping to navigate the chaos within.

Cross-legged on the floor, he closes his eyes and begins his breathing exercises, each inhalation a thread pulling him toward clarity. The room fades away, replaced by a void where thoughts drift aimlessly. He searches for the source of his turmoil, the cosmic frequency that now shapes his consciousness. His mind's eye conjures geometric patterns, fractals folding and unfolding, each one revealing a deeper layer of existence.

Time becomes a malleable concept, moments expanding and contracting. He surrenders to the flow, seeking to understand the overwhelming knowledge that the equation embodies. Emotional waves crash over him—fear, regret, and a burgeoning resolve to make amends. He envisions the seekers, their minds unraveling, each one a mirror reflecting his own fragmented consciousness. Their anguish intertwines with his, creating a tapestry of collective suffering.

As he delves deeper, dawn begins to break outside, the first rays of light seeping through the cracks in the curtains. This nascent glow brings a sense of awakening, a metaphorical reminder of the clarity he seeks. The light, soft but insistent, nudges him toward an epiphany.

He opens his eyes slowly, the room bathed in a gentle luminescence. The chaos within him recedes slightly, replaced by a burgeoning sense of responsibility. He understands now—this equation, this portal to uncharted dimensions, is not merely a tool but a profound reflection of the human psyche. It magnifies the internal state of its interpreter, amplifying both wisdom and madness.

Jaxon rises, moving to the window. He looks out at the city below, the streets slowly coming to life with the new day. He feels the weight of his discovery pressing down on him like a leaden shroud, but within that pressure lies a spark of determination. He realizes that the path forward is not about unlocking more knowledge but about guiding those who seek it, helping them navigate the labyrinth of consciousness without spiraling into chaos.

He knows he must take action, to protect the seekers from themselves and to mitigate the damage already done. His mind settles on Angelica once more, her presence a beacon in his stormy sea of thoughts. Perhaps mending his relationship with her could be the first step in healing the breach his equation has created.

With renewed resolve, Jaxon begins to plan his next moves. He will reach out to Angelica, try to bridge the chasm between them. She is his tether to the human experience he so desperately needs to ground himself. As for the seekers, he will approach them with caution, offering guidance and wisdom instead of unfettered access to the equation's depths.

As he stands there, watching the sunrise paint the sky in hues of hope, a sense of peace starts to settle within him. Jaxon understands that the journey ahead will be fraught with challenges and that he must bear the burden of his creation. But in this moment of dawning clarity, he feels ready to face the responsibilities before him, to become the guardian of both the knowledge he's uncovered and the fragile minds it has touched.

The path forward is uncertain, but for the first time since the lightning strike, Jaxon feels a flicker of hope. He will take each step with mindfulness and care, knowing that his actions could either mend the frayed fabric of reality or tear it further apart. And so, with the light of a new day illuminating his thoughts, he sets his intention to move forward with purpose and compassion.

Chapter 13

DANCES WITH THE DIVINE

Jaxon settles into the empty room, the natural light pouring through the tall windows, casting soft, almost ethereal rays across the polished wooden floor. The room feels like a sanctuary, a stark contrast to the turmoil that has typically engulfed his life. He closes his eyes, inhaling deeply and exhaling slowly, each breath a deliberate attempt to pare away layers of lingering tension.

Now deeply focused on his breath, he begins to attune himself to the rhythmic sounds of nature just outside. Birds chirp lightly in the distance, their melodies blending with the whisper of leaves rustled by a gentle breeze. The scent of fresh earth and greenery filters in through the open windows, and Jaxon allows these sensory inputs to lull his body into a state of relaxation. Memories of crowded kitchens and the cacophony of bustling urban streets fade, replaced by the harmonious cadence of his natural surroundings.

As his breath ebbs and flows, Jaxon visualizes colors appearing in his mind's eye—soft pastels at first, then vibrant hues swirling in kaleidoscopic patterns. These colors dance and merge, forming shapes that pulse with a life of their own. With each breath out, he feels his thoughts dissolving, the clutter of his past, the pain of addiction, the guilt of fractured relationships, melting into the vivid depictions before him. He remembers the metallic clang of pots and pans, the searing heat of the kitchen, and the realization that each dish he created was a microcosm of his internal chaos.

Jaxon sinks deeper, sensing a profound connection to the earth beneath him. It is as though his very molecules resonate with the sensation of being grounded, each one vibrating softly, attuned to an ancient rhythm. The floor beneath him feels almost like a pulsating entity, alive with potential. He imagines roots extending from his body, intertwining with the soil, drawing sustenance and strength from it. In this moment, he is not merely a man but an integral part of the world's living tapestry.

Thoughts of his sister, Angelica, shadow his peaceful mind for a moment. He recalls her face, the concern etched into her features, reflecting the emotional chasm that has widened between them. His fall into addiction had driven a wedge into their relationship, leaving behind scars that both of them carried. The weight of his failures presses down on him, but he counteracts it with intentional breaths, allowing the earth's energy to percolate through him.

Colors fade into the background, replaced by an even deeper void of inner stillness. This void is not empty, but full of potential—every breath a reverberation through this vast, uncharted space within him. Jaxon recognizes this as the threshold of his consciousness, the point where the individual blends seamlessly with the universal. He feels the enormity of the universe, yet in this stillness, there is no fear, only an intoxicating calm that promises deeper revelations.

As he reaches this depth, he recalls his culinary mentor, whose sudden death had sent Jaxon spiraling into this abyss. The anguish tethered to that loss had been a driving force towards his past decisions, yet, here in this tranquil state, he feels a softening—a possibility of reconciliation with that pain. The trauma that once seemed insurmountable feels slightly less mountainous, as if his own inner stillness begins to erode its sharpest edges.

Finally, Jaxon's thoughts crystallize around the notion of interconnectedness, a recurring theme in his recent experiences. He ponders the lives he has touched, both in his former career as a chef and in his new, uncertain path as a cosmic seeker. The fear of losing himself in the vastness of consciousness confronts him, but he balances it with the comforting realization that every individual thread contributes to the larger weave of existence.

Jaxon now dwells in a profound state of tranquility, his mind a quiet surface upon which the universe whispers its secrets. The room, bathed in natural light, feels both expansive and intimate, a silent witness to the transformative journey about to unfold. Here, on the cusp of cosmic revelations, Jaxon feels a burgeoning of hopeful potential–a sign that perhaps, redemption and self-discovery are within his grasp. The stage is set for him to delve deeper into the mysteries that beckon from within.

As he continues to sit in the stillness, memories begin to surface–memories he has long tried to suppress. The image of his parents flashes across his mind. Their disapproving glances, the tense silences at family dinners, the palpable disappointment in their eyes as his life unraveled. He remembers the arguments, the accusations, the sharp words that cut deeper than any knife. His parents, once a source of comfort and guidance, had become reminders of his failures, and eventually, they too distanced themselves, unable to watch their son destroy himself.

The guilt of those broken relationships weighs heavily on Jaxon. He knows that his path to redemption isn't just about understanding the universe or helping others find their way. It's also about facing the ghosts of his past and seeking to heal the wounds he has inflicted on those who once loved him unconditionally. The realization brings a new sense of urgency to his quest. It's not just about finding enlightenment; it's about making amends, about seeking forgiveness from those he has hurt.

The thoughts swirl around in his mind, creating a storm of emotions. He feels the pull of the past, the weight of unresolved issues, and the fear of confronting them head-on. But as he sits there, rooted in the present, he realizes that he can't move forward without first addressing the shadows that lurk in the corners of his mind. The journey ahead is not just about cosmic understanding; it's about personal reconciliation, about finding peace with himself and those he has wronged.

As the light in the room shifts, indicating the passage of time, Jaxon slowly opens his eyes. The room feels different now, charged with a sense of purpose. He knows what he must do. The path ahead is clearer, though fraught with challenges. But for the first time in a long while, Jaxon feels ready to face them, armed with the knowledge that true enlightenment isn't just about the mind—it's also about the heart.

Jaxon sits cross-legged in the room, bathed in natural light streaming through large windows that overlook a serene garden. His eyes are half-closed, his breath steady and rhythmic. The air hums softly with the sound of chirping birds and the distant trickle of a small fountain. For a moment, the world outside falls away, leaving Jaxon suspended in a cocoon of silence.

Within the depths of his consciousness, colors begin to unfurl—vivid hues of blue and green, punctuated by vibrant reds and oranges. These colors swirl and morph into intricate geometrical designs. Jaxon's mind envisions patterns that resemble ancient mandalas, each line and curve flowing seamlessly into the next. Triangles, circles, and hexagons dance before him, creating a tapestry of interconnected forms that pulse with life.

As the patterns intensify, Jaxon contemplates their significance. He feels a resonance with each shape, understanding that these fractals represent the very fabric of existence. His thoughts drift to the mathematical equation he has been developing, and he sees how each symmetrical design echoes the principles of sacred geometry he has embedded in his work. This equation is no longer just a formula; it's a map, a cosmic blueprint that transcends the limitations of his previous understanding.

Jaxon recalls his early days in the kitchen, where the art of cooking was an alchemy of flavors and textures. The meticulous precision required to balance ingredients now mirrors the precision of the geometric designs swirling before him. The discipline that once brought culinary masterpieces into existence now channels his understanding of the universe. Each fractal, each equation, intertwines with the passion he once had for culinary arts, transforming it into a cosmic exploration of reality.

A sudden surge of energy courses through him, breaking his reverie. He opens his eyes, half expecting the room to remain unchanged, but finds it shimmering with cosmic light. The walls vibrate softly, absorbing and refracting the light into a kaleidoscope of luminescent threads connecting every corner of his sanctuary.

It's as if the patterns he saw within his mind have bled into the physical space, transforming it into a multidimensional canvas.

Jaxon's heart pounds with a mix of joy and fear. The intensity of this experience is almost overwhelming, yet he feels an innate understanding that this is a reinforcement of his place in the cosmic order. His mind flashes to memories of his addiction, the darkness that engulfed him and kept him from this clarity. It's a stark contrast to the brilliance he now witnesses, a reminder of how far he has journeyed.

In this moment of enlightenment, his mind wanders to his estranged sister, Angelica. He remembers their shared childhood, the bond that once brought light into his darkest hours. The geometric designs now represent more than just mathematical beauty—they symbolize his desire to reconnect, to reach out and mend their fractured relationship. This newfound understanding between the cosmic and the personal pushes him further into the realms of forgiveness and redemption.

Jaxon embraces the sensations surrounding him. His heart brims with newfound strength and wisdom, each pulse of cosmic energy reinforcing his connection to the universe's vastness. He no longer feels detached; he is a part of the intricate web that spans dimensions, thoughts, and emotions. His equation is not just an intellectual pursuit but a bridge to a deeper, more profound unity that transcends the physical.

He accepts the light and energy enveloping him, allowing them to cleanse his fears and doubts. The room, once a simple meditative space, now pulsates as though breathing with him. Jaxon understands that he has unlocked another layer of existence, one that aligns him closely with the principles of interconnectedness that he now grasps more fully. This realization is both humbling and empowering, guiding him toward a destiny intertwined with the cosmic fabric.

Culinary creativity once expressed through flavors is now a channel for cosmic insights. He perceives the culinary studio he envisions as a sanctuary where seekers can taste the patterns of the universe in each bite. This is his next journey—using food as a medium for others to explore their own paths to enlightenment.

Jaxon's eyes close once more, and he breathes deeply, feeling an immense clarity and connection. The fractals continue their dance, each turn and twist reflecting the vast and infinite possibilities that lie ahead. His heart beats in rhythm with the universe, and he knows that he is ready to share this illuminated path with others, guiding them toward their own moments of cosmic realization.

The sensation of interconnectedness lingers as Jaxon slowly returns from his meditative state. The room, with its gentle light and soothing ambiance, feels like a cocoon of serenity. Yet, within that peace, there is a stirring–a sense of urgency, a recognition that his journey is not one he can undertake alone. The visions of cosmic patterns, the swirling fractals, have given him a glimpse of a greater purpose, but they have also revealed the isolation he has imposed on himself.

Jaxon realizes that his path to enlightenment, while deeply personal, is also inherently communal. The insights he has gained are not just for his own understanding; they are meant to be shared, to be explored in concert with others who seek the same truths. The culinary studio he envisions is not just a place of creation–it is a place of connection, where individuals can come together to explore the mysteries of existence through the universal language of food.

The thought of opening himself up to others, of sharing his journey and his insights, fills Jaxon with both excitement and trepidation. He knows that the path ahead will not be easy. There will be challenges, both external and internal, as he navigates the complexities of human relationships and the intricacies of cosmic understanding. But he also knows that he cannot shy away from these challenges. To do so would be to deny the very essence of the interconnectedness he has come to understand.

Jaxon rises from his seat, feeling a sense of purpose infuse his every movement. He knows that the time has come to take the next step in his journey–to reach out, to reconnect, and to begin the process of building the culinary studio that will serve as a bridge between the physical and the metaphysical. As he moves through the room, gathering his thoughts and his belongings, he feels a sense of anticipation, a quiet excitement that builds within him like a rising tide.

He takes a moment to pause by the window, gazing out at the garden beyond. The world outside is alive with possibility, a reflection of the infinite potential that lies within each of us. Jaxon knows that he has been given a gift–a glimpse into the workings of the universe, a deeper understanding of the interconnectedness that binds us all. And with that gift comes a responsibility to share it, to use it to help others find their own path to enlightenment.

The sunlight bathes him in warmth as he stands there, and Jaxon feels a deep sense of gratitude–for the journey that has brought him to this point, for the insights he has gained, and for the opportunity to share those insights with others. He knows that the path ahead will be challenging, but he also knows that he is not alone. The universe is with him, guiding him, supporting him, as he takes this next step in his journey.

With a final deep breath, Jaxon turns away from the window and moves towards the door. He is ready to face whatever challenges lie ahead, ready to share his journey, his insights, and his love for the culinary arts with the world. The road to enlightenment is long and winding, but Jaxon is prepared to walk it, one step at a time, knowing that each step brings him closer to the unity and understanding he seeks.

As Jaxon sinks deeper into his meditation, the room around him fades into an ambient glow. He feels a gentle warmth emanating from the floor beneath him, an anchoring force that roots him to this precise moment in time. The natural light streaming through the windows takes on an ethereal quality, casting dappled patterns that dance on the wooden floor.

Within his mind's eye, webs of light begin to form, intricate and dazzling, stretching across an infinite canvas. These glowing filaments pulse with life, connecting nodes that seem to represent individual thoughts, energies, and souls. Each thread vibrates with color and electricity, creating an endless tapestry of interconnectedness. Jaxon feels a profound sense of belonging, a nostalgic echo of unity he has long yearned for.

Phrases and ideas begin to swirl within him, most prominently, "as above, so below." This ancient axiom reverberates through his consciousness, expanding to fill every corner of his mind. He contemplates its essence: the microcosm mirroring the macrocosm, the boundless universe reflecting the intricate patterns within each person. The vision solidifies, metaphorically stitching together the vastness of existence with the minutiae of daily life.

The fractal equation he has constructed emerges in his mind's eye, not as a sterile mathematical formula, but as a living, breathing symbol of interconnectedness. He perceives its true purpose: a bridge linking individual human experiences to a collective consciousness that spans across dimensions. This revelation brings a blend of pride and trepidation–his creation is powerful, but he fears its misuse by those unprepared for its depths.

Jaxon resolves with newfound clarity that his role extends beyond the realm of pure mathematics. His culinary artistry takes on a transformational significance. Each dish he intends to create must now serve as a gateway, a sensory experience embodying the unity he has come to understand. He envisions his culinary studio not merely as a place to prepare and share food but as a sanctuary–a sacred space for seekers to embark on their own voyages of self-discovery.

In his mind, he begins crafting an array of dishes infused with geometric patterns and cosmic motifs. He sees plates adorned with exquisite presentations, where every ingredient symbolizes a deeper meaning–organic, fractal shapes that resonate with the harmony of the universe. His cooking is no longer just about taste; it becomes an intricate dance of flavors that guides others toward cosmic insights.

Feeling the weight of this responsibility, Jaxon embraces his destiny. The visions of interconnectedness translate into a tactile determination; he feels a magnetic pull urging him to act, to transform his revelations into tangible forms. A sense of unity and purpose courses through him, merging his once fragmented existence into a cohesive whole. With a deep breath, he understands the impact of his every action within the vast web of life, ready to share his newfound enlightenment through the universal language of food.

The sense of urgency that has been building within him reaches a crescendo. Jaxon knows that he cannot wait any longer. The time for contemplation has passed; now is the time for action. He must begin the process of bringing his vision to life, of creating the culinary studio that will serve as a bridge between the physical and the metaphysical, between the individual and the collective, between the self and the universe.

Jaxon rises from his seated position, feeling the energy coursing through his veins. His mind is alive with ideas, each one more vibrant and compelling than the last. He knows that the road ahead will be challenging, but he also knows that he is ready to face those challenges head-on. The insights he has gained, the understanding he has achieved, have given him the tools he needs to succeed.

He moves through the room with purpose, gathering the materials he will need to begin the process of creating his culinary studio. His hands move with a practiced ease, the muscle memory of his years as a chef guiding him as he begins to sketch out plans, to write down recipes, to envision the space that will become a sanctuary for those seeking enlightenment through the sensory experience of food.

As he works, the room around him seems to come alive, the walls vibrating with the energy of creation. The light streaming through the windows takes on a golden hue, bathing the room in a warm, comforting glow. Jaxon feels a sense of peace, of contentment, as he pours his heart and soul into the task at hand. He knows that this is what he was meant to do, that this is his true calling.

Hours pass in a blur of activity, and by the time Jaxon finally pauses to take a breath, the room is filled with the fruits of his labor. Sketches of the studio adorn the walls, recipes are scattered across the floor, and the air is thick with the scent of fresh herbs and spices. Jaxon takes a step back, surveying the scene with a sense of satisfaction. He has taken the first step on a journey that will change his life, and the lives of those who seek his guidance, forever.

But even as he stands there, basking in the glow of his accomplishments, Jaxon knows that this is only the beginning. The road ahead is long and winding, and there will be many obstacles to overcome. But he is ready for the challenge. He is ready to embrace his destiny, to become the bridge between the physical and the metaphysical, to guide others on their own journeys to enlightenment.

With a final deep breath, Jaxon turns away from his work and moves toward the door. The sun has set, casting the room in a warm, golden light. As he steps outside, he feels the cool evening air wash over him, invigorating him, filling him with a renewed sense of purpose. The world outside is quiet, peaceful, and Jaxon knows that he is exactly where he is meant to be.

He takes a moment to look up at the sky, the stars twinkling above him like beacons of hope. The universe is vast, infinite, and Jaxon knows that he is but a small part of it. But he also knows that every action he takes, every decision he makes, ripples out into the cosmos, affecting everything around him. The interconnectedness he has come to understand is not just a concept; it is a reality, a truth that guides his every move.

Jaxon begins to walk, his steps sure and steady, the path ahead illuminated by the light of the stars. He knows that the journey will be difficult, that there will be moments of doubt and fear. But he also knows that he is not alone. The universe is with him, guiding him, supporting him, as he takes each step toward his destiny.

The road stretches out before him, winding through the darkness, but Jaxon is not afraid. He walks with purpose, with confidence, with the knowledge that he is exactly where he is meant to be. The culinary studio awaits, a sanctuary of enlightenment and understanding, and Jaxon is ready to bring it to life.

The meditative trance releases its hold, and Jaxon ascends from the depths of his consciousness, a profound clarity settling upon him like a second skin. He can almost hear the hum of the cosmos as his awareness returns to the serenely lit room, the golden shafts of sunlight warming the wooden floor and wrapping him in a cocoon of tranquility. His beating heart finds its rhythm, synchronized with the natural cadence whispering through the leaves swaying just beyond the window.

Jaxon kneels to the blank pages of his journal, the crisp scent of paper mingling with the soft aroma of sandalwood lingering from a nearby burning stick of incense. With a hand both calloused and tender, he starts writing, each stroke of the pen a deliberate dance, capturing the fragments of enlightenment that shimmer at the edge of his consciousness. The interconnectedness of all things becomes his muse. He sketches geometric patterns, fractal designs spiraling outwards, each line resonating with the newly uncovered truths seeping from his mind. Words flow from him, a stream of cosmic wisdom blending seamlessly with his personal reflections, forming a narrative tapestry meant for those on the cusp of their own revelations.

His thoughts drift towards the culinary studio, born from both necessity and visionary promise. He imagines transforming the empty space into a sanctuary, a realm where seekers of truth could embark on their journeys through the senses. The studio becomes alive in his mind's eye—marble countertops gleaming under the soft glow of pendant lights, shelves brimming with jars of vibrant spices and aromatic herbs, each waiting to become part of a symphonic experience designed to awaken dormant consciousness.

In this haven, Jaxon pictures guiding hands, steady and nurturing, teaching the art of infusing every dish with a piece of the infinite. He envisions dishes as reflections of the cosmos, each ingredient a star, each flavor a note in the melody of the universe. He sees seekers entering the studio burdened by the chaos of the outside world, leaving touched by the alchemy of taste and enlightenment.

A smile breaks across his face, a smile unburdened by the ghosts of his addiction, for now he understands. His journey of self-destruction has not been in vain; every fall, every moment of despair now metamorphoses into a step towards redemption. The pain that once stitched his reality together now unravels, threads of wisdom in their place.

He closes the journal, a sense of profound purpose enveloping him. This is no longer about linear paths but about spaces filled with love, places sculpted from the same divine patterns he has come to understand. The culinary studio is his future, and through it, a portal for others. Lost souls like him, broken by their battles, will find guidance not just in equations and meditations but in the shared experience of creation and nourishment.

His mind revisits past relationships, the ones strained and tainted by his addiction. He reflects on his estranged sister, Angelica, and their fractured connection. The culinary studio isn't just for strangers; it is a space that might one day mend the bond with his sister, cooking together like they once did in their grandmother's kitchen, flavors melding seamlessly, serving as metaphors for healing.

Destiny now flows through him with a force unstoppable. With each culinary creation, every dish plated with deliberate intention, Jaxon intends to weave these fractal patterns into the lives of seekers. As he ponders the infinite nuances of flavor and spiritual sustenance, he notes the sacred geometry that ties taste to soul. This is his mission: to feed more than the body– to nourish the spirit.

Jaxon organizes his notes meticulously, the fledgling plans for his culinary studio taking form. His eyes scan the room, now a nexus of intersecting energies, the walls still echoing with cosmic light. His culinary artistry will reflect the profoundness of his meditative revelations, each dish a step towards enlightenment.

As he prepares to close this chapter of introspection, he relishes the calm before the tempest of creation. The studio promises to be a nexus where culinary skill and cosmic wisdom converge, providing a beacon for the lost–each bite a revelation, each flavor a universe waiting to be explored. His soul vibrates with the possibilities that lie ahead, each plate of food another fractal in the infinite expanse of existence.

Jaxon can already taste it–hope mingled with redemption, a future where his culinary studio stands as a sanctuary for those seeking the same depth of understanding that saved him.

Jaxon's heart swells with anticipation as he begins to plan the first steps towards bringing his vision to life. He knows that the culinary studio will be more than just a place to cook; it will be a sanctuary for those seeking not only physical nourishment but also spiritual and emotional healing. Every detail, from the layout of the kitchen to the selection of ingredients, will be infused with the principles of interconnectedness and sacred geometry that have become the foundation of his new understanding.

He envisions a space that is both functional and beautiful, where the act of preparing food becomes a meditative practice, a way to connect with the deeper truths of the universe. The kitchen will be a place of calm and focus, where the process of cooking is as important as the final dish. Each meal will be an opportunity to explore the connections between the self and the cosmos, to taste the patterns of the universe in every bite.

As Jaxon begins to sketch out the layout of the studio, he feels a sense of excitement building within him. He imagines the different stations–one for chopping and prepping ingredients, another for blending spices, and yet another for plating and presentation. Each station will be designed to encourage mindfulness, with tools and ingredients arranged in a way that promotes a sense of flow and harmony.

He sees the studio as a place where people can come to reconnect with their senses, to experience the joy of creating something beautiful and nourishing with their own hands. It will be a space where the act of cooking becomes a form of meditation, a way to quiet the mind and focus on the present moment. Through the preparation and sharing of food, Jaxon hopes to help others find the same sense of peace and clarity that he has discovered on his own journey.

As he continues to plan, Jaxon's thoughts turn to the people he hopes to help through his studio. He envisions seekers of all kinds– those struggling with addiction, those grappling with loss, those searching for meaning in their lives–finding solace and inspiration in the act of cooking. The studio will be a place of healing, where the simple act of preparing and sharing a meal becomes a path to self-discovery and spiritual growth.

Jaxon knows that the road ahead will be challenging. There will be practical obstacles to overcome–finding the right space, gathering the necessary resources, and building a community of like-minded individuals who share his vision. But he also knows that he is not alone. The universe is with him, guiding him, supporting him, as he takes each step towards bringing his dream to life.

With a renewed sense of purpose, Jaxon begins to make a list of everything he will need to bring his vision to fruition. He writes down ideas for classes and workshops, each one designed to teach the principles of mindful cooking and the connections between food, the self, and the cosmos. He envisions guest instructors– chefs, healers, and spiritual teachers–who will share their own insights and expertise, adding depth and diversity to the offerings at the studio.

He also begins to think about the people he wants to invite into this space. Old friends and colleagues from his days as a chef, people he has met on his journey of self-discovery, and even those he has lost touch with over the years. Jaxon feels a deep desire to reconnect with the people who have been part of his life, to share with them the insights and understanding he has gained.

One person in particular comes to mind–his sister, Angelica. Their relationship has been strained for so long, burdened by the weight of his addiction and the pain it has caused. But Jaxon feels a glimmer of hope that the culinary studio could be a place where they can begin to heal the rift between them. He imagines inviting her to the studio, showing her the space he has created, and sharing a meal together as they begin the process of rebuilding their bond.

The thought of reconnecting with Angelica brings a sense of warmth and hope to Jaxon's heart. He knows that it will take time and effort to mend their relationship, but he is willing to do whatever it takes. The culinary studio is not just a place for strangers to find healing–it is also a place for Jaxon to heal the wounds of his own past, to make amends with those he has hurt, and to build a new life based on the principles of love, connection, and understanding.

As Jaxon continues to plan, the sun begins to rise, casting a warm, golden light across the room. The new day brings with it a sense of possibility, a reminder that the future is not set in stone but is shaped by the choices we make in the present. Jaxon feels a deep sense of gratitude for the journey that has brought him to this point, for the insights he has gained, and for the opportunity to share those insights with others.

With the first rays of sunlight filling the room, Jaxon takes a deep breath and closes his journal. He is ready to take the next step, to begin the process of bringing his vision to life. The culinary studio awaits, a sanctuary of healing and enlightenment, and Jaxon is ready to embrace his role as its creator and guide.

As he steps outside into the crisp morning air, Jaxon feels a sense of peace and contentment that he has not felt in years. The world around him is alive with possibility, and he knows that he is exactly where he is meant to be. The road ahead will not be easy, but Jaxon is ready for the challenge. He is ready to share his journey, his insights, and his love for the culinary arts with the world. And with each step he takes, he knows that he is one step closer to the unity and understanding he has sought for so long.

The culinary studio, once just a vision in his mind, is now a tangible reality, a place where Jaxon will help others find their own path to enlightenment. And as he walks towards the future, Jaxon knows that he is not just creating a space for others–he is also creating a space for himself, a place where he can continue to grow, to learn, and to heal.

Chapter 14

THE FIRE WITHIN

Jaxon sits alone in his dimly lit living room, surrounded by a chaotic mess of papers, scribbled equations, and remnants of his culinary past. The flickering light bulb overhead adds an uneasy ambiance to the room. Plates with dried sauce, abandoned for days, sit among empty bottles and half-burned candles. The profound silence is interrupted only by the occasional distant honking of cars outside, a reminder of the world still spinning on, ignorant of his turmoil.

A shadow shifts at the doorway, and Carla bursts in, her face a mix of concern and anger. She scans the disarray with widened eyes before they lock onto Jaxon, who doesn't even flinch.

"Jaxon, this is insane," she exclaims, her voice trembling with emotion. "Do you even realize what you've done? People are losing their minds! The seekers... they're broken, Jaxon."

Her words hang in the air like an impending storm, each syllable laden with urgency. Carla is visibly shaken, her presence a stark contrast to the serenity she once embodied as his confidante. She steps through the room, navigating the clutter, as if the chaos itself were a physical manifestation of the equation's effects.

A television in the corner hums to life, replaying news segments. Headlines scroll across the screen, each one more damning than the last: "FRACTAL EQUATION FUELS MENTAL EPIDEMIC," "USERS SUFFER SCHIZOPHRENIC EPISODES," and "SCIENTIFIC COMMUNITY IN UPROAR." Faces of distressed individuals appear next, testimonials of those whose lives have disintegrated into fractals of their former selves.

Jaxon feels the weight of each story, each statistic. The guilt is a molten core within him, threatening to erupt. But there's also a flicker of pride–an acknowledgment of the genius that facilitated such profound, albeit tragic, results.

The door swings open again, and Dr. Miles rushes in, drawing an immediate line of tension in the room. He is visibly weary, the toll of relentless pressure from the scientific community evident in his furrowed brow and tired eyes.

"Carla, wait," he pants, as if he had been running. "You can't just dismiss Jaxon's work like that."

Miles' tone shifts to a mix of frustration and desperation. "Jaxon, your equation is... it's a breakthrough. We just need to understand it better, control it. This backlash... it's just fear of the unknown."

Jaxon can only watch, absorbing the fiery exchange between two significant yet opposing forces in his life. Carla stands her ground, her passion uncontainable. "Fear of the unknown? Miles, people are shattered! Just look around you. Look at Jaxon!"

Their words clash, a tempest of rationale against emotion. Miles rubs his temples, his skepticism battling with his ambition. He respects Jaxon, but the urgency to validate or debunk the equation surges within him. The new blurbs flashing on the screen unexpectedly meld into his psyche, integrating layers of ego and dread–components of every thought, fractalized within the equation's birthright.

A gust of wind whips through the room as Allegra storms in, her entrance like a mystical tornado. Her eyes burn with fervor, and her movements are fluid, almost otherworldly, as she steps into the chaotic fray.

"Jaxon," she breathes, her voice a captivating cadence. "This isn't about fear or science. This is about enlightenment, about spiritual transcendence!"

She extends her arms, reaching out to Jaxon with impassioned plea. "Join me, Jaxon. Together, we can guide these seekers to a higher plane of existence. They just need the right guidance."

Her proposition saturates the air, pushing the room's atmosphere into a cacophony of clashing drives–empirical knowledge, human concern, and spiritual awakening. Jaxon's mind spins, caught in the tempest of these conflicting ideologies. He catches glimpses of desperation in Carla's eyes, fervor in Allegra's, and a mix of ambition and frustration in Miles'. He is a crucible, bearing the weight of their expectations, their fears, and their faith.

Jaxon begins to quake inwardly. Each character's desires weigh down on him, threatening to crush his already fragile psyche. The room, once his sanctuary, now feels like a claustrophobic cage, closing in on him, piece by cluttered piece.

"Stop, just stop," Jaxon finally mutters, his voice barely audible yet laden with the chaos in his mind.

The room falls silent, but the reprieve is short-lived as the walls seem to close in further, magnifying his turmoil. The dissonance within swells, each voice pulling him in different directions. It's too much.

He sinks deeper into his chair, overwhelmed, unable to escape the torrent of conflicting forces.

Jaxon trudges through the sterile corridors of the mental health facility, the glaring white walls and antiseptic scent almost suffocating. Each step he takes echoes hollowly, amplifying the unnatural quiet that hangs heavily in the air. His heart tightens with every faint sob or detached laugh that filters out from the rooms he passes. The faces he sees through small, rectangular windows haunt him–gaunt, pale figures, staring vacantly or muttering incoherently. The vibrant chaos of his former life seems a distant memory, swallowed by the sterile gloom around him.

As he approaches the nurse's station, a young woman with tired eyes looks up from her paperwork. She offers a sympathetic smile, but the corners of her mouth twitch with worry. Jaxon forces himself to speak, his voice cracking under the weight of his guilt.

"How is he?" he asks, knowing that even the words feel inadequate. The nurse glances at a folder before her, flipping through pages filled with charts and notes. "He's detached, experiences vivid hallucinations. He speaks in equations and patterns, as if he's caught in an endless loop. We've sedated him for his own safety, but it's... it's severe."

Jaxon swallows hard, his throat tightening. Each word feels like another blow, another reminder of the devastation his equation has wrought. The fragmented lives, now reduced to patterns and fractals, mirror the shattered state of his own mind. He remembers his time in the depths of addiction, when reality itself seemed a twisted, unending maze. This is different, yet agonizingly familiar.

In one of the rooms, he spots a disoriented seeker sitting on the floor, rocking back and forth, whispering fragments of his equation. Jaxon's stomach churns, and his legs feel like lead, but he forces himself to approach. The seeker, a young man with hollow eyes, gazes up as Jaxon steps inside. Recognition flashes in his eyes, followed swiftly by a torrent of emotion.

"You!" The seeker's voice rises, desperation tinging his words. "Why did you– Why did this happen to me? I... I just wanted to understand!"

"I'm sorry," Jaxon whispers, his voice barely audible. "I'm so sorry."

The seeker's outburst turns to a pleading sob. "Fix it. Please, fix me."

Tears sting Jaxon's eyes as he sinks to his knees, reaching out, though he stops short of touching. What could he say or do that would mend the unraveling mind before him? Each heartbeat feels like a hammer against his chest, echoing with the reminder of his profound failure. He remembers the vibrant streets, the scents of spices and herbs, the laughter, and the chaos–it feels like another lifetime.

Driven by a mixture of empathy, remorse, and helplessness, Jaxon shakily rises and exits the room. The guilt weighing him down is almost palpable, threatening to crush his spirit beneath its relentless pressure. The sights and sounds of the facility– once faceless names, now devastatingly real–stay with him as he makes his way through the labyrinthine corridors.

Stepping outside, the cold air hits him like a slap, but it offers no solace. He stands at the entrance, staring at the concrete beneath his feet, lost in thought. The faces of the seekers, their brokenness, their pleas for help, intertwine with his memories, creating an unbearable quilt of suffering and despair. The smell of antiseptic still clings to him, a stark reminder of the sterile reality he has forged for others–and for himself.

"How did it come to this?" he thinks, gripping the straps of his bag. He is assailed by visions of his culinary days, where each dish was a piece of art designed to delight the senses, not twist the mind. Back then, his creations were meant to connect people, to bring joy and exploration through flavor and experience. Here, in this somber world of sterile walls and detached souls, he witnesses the opposite: a grim parody of enlightenment that disjointed minds rather than unified them. His life of vibrant chaos, once so full of possibility and life, now feels like a cruel joke

He looks up at the sky, a vast gray canvas devoid of the fractal patterns that once consumed his vision. Jaxon feels the weight of his pursuit for knowledge and meaning, now tangled with a sorrow so profound that it threatens to tear him apart. Every facet of his being screams for redemption, for a way to make things right, but he is lost in the endless fractals of his conscience. The heavy doors of the facility close behind him with a dull thud, sealing the cries and whispers of the seekers within.

Standing alone, the reality of the situation crashes over him in waves. He understands now, with a clarity that cuts like a knife–his relentless pursuit has consequences, not only for himself but for the myriad minds he has unwittingly fragmented. In the cold, unforgiving light of day, he recognizes his role in the cosmic equation–a fractal once thought beautiful, now a web of suffering that ensnares them all.

Eyes fixed and unseeing, he knows he must confront the truth head-on, bear the burden of his genius, and find a way to bring harmony back to the chaos he has unleashed. As he steps away from the facility, each step is heavy with resolve, and he knows that the path ahead will be fraught with challenges, but he must embark on it.

The clinking of the bottle against the rim of the glass echoes through Jaxon's cluttered home, a hollow sound that mingles with the oppressive silence. He pours a generous amount of amber liquid, its scent a potent mix of comfort and regret. The burn of the first sip provides a fleeting numbing warmth, a temporary barrier against the tumultuous storm brewing inside him. Shadows dance on the walls, cast by the dim, flickering light of a solitary lamp, creating an atmosphere thick with tension and unspoken anguish. Jaxon's eyes glaze over as he nurses his drink, his mind drifting into the labyrinthine corridors of memory. Vivid flashbacks assault him, images of seekers once full of hope, now hollow husks of their former selves. Among them, a young man, eyes widening with wonder as he first grasped the equation, now sits in a sterile facility, rocking back and forth, eyes distant, lost. Jaxon sees an older woman who had approached the equation with reverence, now muttering incoherently, trapped in a reality he inadvertently created. Their pain throbs within him, each heartbeat a reminder of his unintentional cruelty.

He clenches his glass tighter, knuckles white, as a surge of guilt and sorrow overtakes him. The physical space around him seems to close in, the walls of his once beloved home–now crammed with equations and culinary remnants–pressing against his chest. The chaotic fusion of his old life of culinary artistry and the new, surreal reality underscores the sense of claustrophobia enveloping his mind.

A shout bursts from Jaxon's lips, raw and primal, reverberating through the room. It is a cry of frustration, anger, and unendurable sorrow that leaks out, unfiltered. His thoughts spiral into a dark vortex, a cacophony of self-reproach and desperate yearning. "Why did I create this?" he screams, voice cracking. "How could I have known... they weren't ready... I wasn't ready!"

He slumps to the floor, the coolness of the wooden planks barely registering through his emotional turbulence. Tears carve silent pathways down his face, mingling with the remnants of his drink. Anguish seeps from every pore, his body a portrait of unrelenting torment. In fits of clarity amid the emotional tempest, he recognizes the magnitude of his actions–not just as a cosmic alchemist but as a flawed human being who must now bear the weight of unintended consequences.

His mind wanders to his past as a chef, the precision and passion with which he once created masterpieces, now overshadowed by this overwhelming burden. How he had once danced through the kitchen, every ingredient a note in a culinary symphony, now seems a distant memory, tarnished by the chaos he has unleashed. The contrast between the vibrant life in his dishes and the desolate outcomes of his equation deepens his regret.

Jaxon's breaths come in ragged gulps as he curls into himself, each inhalation a struggle against the profound weight on his chest. His internal dialogue remains a constant thrum, a relentless self-interrogation blending into semi-coherent acknowledgments of his role in others' suffering. Recognition dawns slowly, painfully–the equation didn't just reveal cosmic truths; it mirrored the inner chaos of those who touched it, amplifying what lay within.

Moments stretch into an agonizing eternity as he lies there, contemplating his next steps. The room is silent now but for the ticking of a clock and the distant hum of the city outside. Jaxon knows the road ahead will be fraught with challenges, but he also understands that he must face the ramifications of his creation head-on.

Exhausted and emotionally spent, he finally acquiesces to the necessity of re-evaluation. The magnitude of his realization crystallizes; enlightenment is not a gift to be recklessly shared but a personal journey each must undertake. Through his pain and regret, a fragile sliver of resolve forms. This is his burden to carry, and his steps moving forward must be measured, thoughtful.

As he lies there, eyes half-open, staring at the shadowed ceiling, Jaxon breathes deeply. The air feels different, thicker–laden with the collective weight of his discoveries and the lives they've touched, both ruined and enlightened. This moment of acceptance marks the beginning of a long, arduous path to redemption.

Jaxon sits perched on the edge of his cluttered desk, his eyes fixed on the ominous pile of papers strewn before him. The chaotic mess of equations and cryptic symbols spirals like a fractal maze over sheets covered in frenzied scrawls. The room, once a sanctuary of culinary creativity, now reflects the fractured state of his mind–the walls, lined with remnants of his past life, vibrant paintings and prestigious awards now symbolize a poignant dissonance.

The silence is punctuated by the sound of his own erratic breathing, his mind racing with the images of those who have been torn apart by his creation. Each geometric pattern on the paper seems to whisper the cries of the seekers–scientists, mystics, and agents—who believed in him only to teeter on the brink of madness. The weight of their desperation bears down on him, compressing his skull in an invisible vise.

He clenches his fists, knuckles whitening. Could they be right? Are Carla's fears justified in asserting the weaponized insanity of his fractal equation? Or are Dr. Miles' logical proofs unveiling promises of unparalleled discoveries? Amidst the turmoil, Allegra's voice resounds, demanding that he harness this cosmic key for spiritual ascendance.

His thoughts flit like fireflies in a dark forest, his consciousness threading through moments etched with their faces. Carla, confronting him with an old friend's shaken resolve; her eyes, once warm, now burning with concern. Dr. Miles, brows furrowed, arguing with the fervor of a scientist at war with his skepticism. Allegra, her mystic gaze piercing through his defenses, imbuing every word with passion and enigma.

As he stands ensnared in this quagmire of conflicted desires, the phone perched on the corner of the desk vibrates incessantly, jarring him back into reality. One frantic call follows another, voices layering over each other in a cacophony of distress. "Jaxon, please, you have to help me," implores one seeker, their voice trembling with desperation. Another, on the brink of sobbing, accuses him of destroying their sanity, "Why did you do this to us?"

Their words crash over him like relentless waves, each call a wrenching cry for salvation, each one a fragment of their unraveling minds. The pressure mounts, suffocating. Their dubious trust metamorphoses into an albatross anchoring Jaxon further into his abyss of guilt. Their despair is a mirror, reflecting his own fracturing psyche.

The room feels smaller, the walls moving closer. The vibrant colors of his past achievements now appear ghostly, phantom remnants of a happier time. Jaxon feels his pulse quicken, the air thickening around him. He stands abruptly, a surge of anger and determination coursing through his veins.

With a resolve born of desperation, he seizes one of the papers from the chaotic pile. His fingers tear through it, each rip resounding with stark finality in the silent room. He throws the torn fragments into the air, watching them flutter like wounded birds, each piece a symbol of his struggle to reclaim control over the chaos he unwittingly unleashed.

As the fragments descend gently, like leaves on an autumn breeze, a tentative sense of empowerment stirs within him. He watches them land, covering the room in shards of what used to contain his grand vision. In each fragment, he sees both his failure and his first step toward redemption. It's a small act of rebellion, but it makes him feel, for the first time in a long while, that he might wrest back authority over his creation.

Exhausted yet steeled, Jaxon stands surrounded by the remnants of his work, contemplating his next course of action. The phone's screen is dark now, silent. He knows he must face Carla, Dr. Miles, and Allegra, clear the air, and navigate the storm of ambitions, fears, and shattered dreams. He must reclaim not just authority over the equation, but over his life, which teeters on the fragile edge of catastrophe.

Chapter 15

ECHOES OF THE ETERNAL

Jaxon stands in his reclusive workshop as the evening shadows deepen around him. His cluttered workspace is illuminated by a single, flickering bulb, casting elongated shadows that dance on the walls. The air is heavy and still, almost as if it is holding its breath, mirroring the weight in his chest.

With a heavy heart, he starts gathering all printed copies of the fractal equation from various nooks and crannies. The papers are strewn about chaotically–some under unfinished culinary blueprints, others wedged between old cookbooks stained with years of culinary experimentation. Each sheet he collects feels like lifting pieces of his own fragmented soul.

As the stack of papers grows, Jaxon's mind drifts back to his past. He sees himself in the bustling kitchens, where his avant-garde dishes once sparked joy and awe. The laughter of satisfied diners rings in his ears, interspersed with memories of late-night creativity fueled by an insatiable desire to push culinary boundaries. But then the images shift, darkening into the haze of addiction. Bottles of alcohol and piles of drugs replace the accolades, and the once-brilliant green in his eyes dims under the weight of regret.

He lays the gathered papers out on a dimly lit table, each sheet a testament to both his genius and his torment. The complicated mathematical notations stare back at him, mocking and mesmerizing. As he stands there, memories flood his mind–not just of the joy these symbols brought him, but of the pain and chaos they unleashed. Each equation is a gateway to realms of unimaginable beauty and peril, a double-edged sword that has the power to enlighten or to destroy.

His hand trembles as he picks up a lighter, its metallic surface cold and unforgiving in his grasp. The flicker of the flame reflects in his eyes, a tiny beacon in the encroaching darkness of his thoughts. Regret and resolve swirl inside him, battling for dominance. He hesitates, his heart pounding, as he tilts the lighter toward the stack of papers. The glow from the flame dances on the intricate symbols, turning them into shadows of their former brilliance.

With a deep breath, he finally sets the flame to the edges of the papers, watching as the fire hungrily consumes the intricate notations. The acrid smell of burning paper fills the room, mingling with the scent of old books and lingering spices. The flames grow, crackling and hissing, as if the very essence of the equations is fighting its own destruction.

As the fire consumes the papers, Jaxon feels a complex blend of sorrow and release. It is as if a part of him, the part that clung to the chaos of his creation, is being set free. Yet, the loss is undeniable–the knowledge he has chosen to destroy is not just scientific, but a piece of his journey, a manifestation of his deepest insights and wounds. It burns away not just the threat it posed, but also the potential for transcendence it held.

The light from the fire dims as the last of the papers turn to ashes, leaving behind a residual smoke that drifts lazily through the air. It curls and sways, casting ethereal patterns that momentarily capture the fractal essence Jaxon sought to control. The smoke lingers, a ghostly presence in the room, reflecting the residual darkness within him–a darkness mingled with the glimmer of hope that perhaps, finally, he is moving toward self-redemption.

The final embers die out, and Jaxon is left in near darkness, the oppressive silence broken only by his own labored breathing. He stands there, staring at the charred remnants of his work, feeling both the weight of his decision and the faint stirrings of a new beginning. The room is filled with the scent of smoke and burnt paper, an olfactory echo of finality.

As the evening slips into night, the workshop grows quieter, the air thick with anticipation for what comes next. Jaxon feels the darkness surround him, but within it, a small, persistent light of determination still flickers, fueling his resolve. With the physical manifestation of his work now gone, he knows he must carry its essence within him, a burden and a gift he alone must bear. The night deepens, but in that darkness, Jaxon stands ready to face the path ahead, scarred but unbroken.

The scene ends with the flames flickering out, leaving a residual smoke in the late evening air, mirroring the darkness Jaxon feels within.

The air is heavy with the smell of burnt paper, a lingering testament to the irrevocable act Jaxon has just committed. He sits in his dimly lit workspace, the silence around him more oppressive than any noise. The empty table in front of him feels like a void, a chasm where the fractal equation once existed. He can't help but feel the weight of this emptiness pressing down on him, much like the burden of his own fractured psyche. Each corner of the room, once vibrant with the energy of frenetic scribbling and breakthrough moments, now feels barren, hollow.

Sitting on the worn-out chair, Jaxon closes his eyes, seeking refuge in the sanctuary of his own mind. He needs to retrieve the concepts and formulas, to make sure they are indelibly etched in his memory. As he begins to summon the sorely missed equations, his thoughts race—memories of what the equations represent flood his senses. They were more than just numbers and patterns; they were a repository of his hopes, his fears, his dreams of redemption. Each twist and turn of the fractal lines mirrored his internal struggle, the battle between genius and madness, clarity and confusion.

He vows internally to carry this knowledge within him, to prevent any misuse. His fingers twitch involuntarily, remembering the physical act of writing those numbers, that intricate dance of pen on paper. Now, everything must be recalled and stored within, safeguarded from the grasping hands of those who would misuse it. The fractal equation is a double-edged sword—capable of immense enlightenment or utter destruction. His responsibility is profound and unforgiving.

Jaxon begins to meditate deeply, his breath slowing, becoming a rhythmic mantra that guides him deeper into the layers of his consciousness. The air around him thickens, charged with the energy of his internal battle. He visualizes the fractal equation, trying to reconstruct it piece by piece. The patterns emerge slowly, not as lines on paper, but as ethereal pathways in his mind, each curve and loop shimmering with meaning.

These patterns are not just mathematical; they are the essence of his experiences. He sees his past failures embedded within them, moments where he fell short, where he gave in to addiction, where he let down those who believed in him. The meditative state brings both clarity and torment. His mind is a crucible, where his shattered thoughts are melted and reformed into something stronger.

Amidst the internal chaos, he focuses on his breathing, grounding himself. For a moment, the weight of responsibility feels unbearable, yet liberating. He is the sole bearer of this arcane knowledge, a gatekeeper to realms beyond human understanding. His shoulders ache under the metaphysical weight, but there is a sense of liberation knowing it is now protected within his mind. The dichotomy of liberation and burden intertwines, a complex dance of freedom and duty.

As Jaxon languishes in this meditative state, memories of his former life as a chef surface. The memories are poignant, imbued with the scent of fresh herbs, the sizzle of ingredients in a pan, the vibrant colors of a meticulously prepared dish. Cooking was once his refuge, a sacred act of creation where he transcended the mundane. Now, his culinary skills hold new meaning–a way to merge his cosmic insights with the tangible world, to transform abstract knowledge into sensory experiences.

Images of mushrooms revealing their fractal gills and caramel swirls forming intricate geometric patterns float into his mind's eye. Cooking, he realizes, is not just about sustenance; it's a form of artistry that can reflect the fractal nature of the universe. The kitchen becomes a metaphoric landscape where he can channel his internal revelations, offering a taste of the infinite to those who partake.

Time loses its meaning in meditation. Hours slip by unnoticed, merging into one surreal moment of enlightenment. When Jaxon finally opens his eyes, night's blanket still envelops his workspace, the moon casting ghostly reflections through the windows. Residual smoke from the fire hovers in the air, adding a dreamlike quality to his surroundings. In this moment, he feels a newfound determination blooming within him, a resolve forged through the crucible of his internal battle.

He knows the path ahead is fraught with trials, but his resolve is unwavering. Carrying the weight of the fractal equation within him, Jaxon understands his role is no longer just that of a creator but a protector. With this knowledge comes a sense of purpose and a willingness to guide others toward their own understanding, be it through culinary artistry or the profound insights of the cosmic patterns he sees.

Jaxon rises from his chair, his mind clearer, his heart fortified. He stands in the residual smoke, feeling both the gravity of his decisions and the promise of what lies ahead. The journey to integrate his extraordinary knowledge with his life's passion has only just begun. He steps out of the shadows of his past, ready to embark on his new path, his resolve as unyielding as the eternal fractal patterns he now holds within.

In the early hours of dawn, Jaxon sits huddled in his introspective space, a labyrinth of unfinished sketches, splattered canvases, and dense mathematical notations. The remnants of his artistic endeavors surround him like relics of a bygone era. Each artifact carries the weight of his complex journey–a path carved through addiction, enlightenment, and now, a painful reckoning. The air is filled with a mix of ink, charcoal, and the faint, lingering scent of burnt paper from the night before.

As he reflects, memories of the seekers he has encountered flood back. Each face appears in his mind's eye, their expressions teetering between awe and agony upon confronting the fractal equation's implications. He recalls the fragile hope that glimmered in their eyes, a hope that often shattered into shards of incomprehension and despair. He thinks of Maya, her intellectual fervor twisting into painful obsession, and of Dr. Miles Carter, whose scientific skepticism barely shielded the existential dread lurking beneath. Allegra Frost's piercing green eyes flash in his memory, her ambition betraying a dangerous edge. They came to him seeking transcendence, but too often, they found only chaos and disarray.

Doubt seeps into Jaxon's consciousness, a creeping vine that entwines around his heart. Was his decision to destroy the physical copies of the equation truly the right choice? The doubt is relentless, shadowing every thought. Could he have guided the seekers better, protected them from the torment that knowledge inflicted? His mind replays the image of the flames consuming the papers, a flickering dance of destruction and liberation. The memory is tinged with the guilt of possibly extinguishing an opportunity for enlightenment. Yet, there remains a sliver of resolve–an understanding that the knowledge is too powerful, too perilous to exist unchecked.

His inner voice whispers, a dialogue that traverses the boundaries of reason and emotion. "You tried," it murmurs, "but was it enough?" The question reverberates, bouncing off the walls of his mind, echoing through the corridors of past decisions. Jaxon's past mistakes loom large, his failures etched into the very fabric of his being. Each misstep stretches into infinity, a fractal of regret.

The potential impact of his knowledge begins to unfold in his thoughts, like an intricate fractal pattern revealing its infinite depths. He envisions the equation as more than mathematical notation; it is a cosmic riddle, a key to dimensions beyond human comprehension. It held the promise of uncharted realms, but only for those prepared for the journey. His reflections are dotted with fragments of memories–seekers overwhelmed by the boundless dimensions, their minds fracturing under the weight of higher-order realities.

Jaxon's musings turn toward the dynamics between him and the seekers. He recalls the initial meetings, the tentative steps toward understanding, and the subsequent unraveling of their psyches. He sees himself not as a gatekeeper, but as an unwilling guardian of a secret too immense for fragile human minds. Each seeker reflected his own fears and aspirations, each falter a mirror to his internal strife.

The realization dawns upon him with the first light of morning, casting elongated shadows and golden hues across his cluttered workspace. He must protect himself and honor the genuine aspirations of seekers who truly yearn for growth. To become a guide rather than a gatekeeper, he must tread the delicate path of responsibility and care, leading by the example of his own journey. He understands now that true enlightenment cannot be handed out like a recipe card; it must be discovered, earned through the crucible of personal experience.

As dawn breaks, the residual smoke from the burnt papers mingles with the crisp, new air of the morning, creating an ephemeral veil that slowly dissipates. Jaxon stands up, the weight of his reflections lifting as a newfound sense of purpose envelops him. His introspective space, once a repository of his past turmoil, now glimmers with the promise of renewal. The remnants of his artistry seem to pulse with latent energy, waiting for a guiding hand.

With a sense of resolve and clarity, Jaxon steps into the dawn, ready to embrace the path of guiding others toward their own illumination, each step a note in the symphony of his redemption.

In the first light of the morning, Jaxon's bright kitchen becomes a sanctuary, its walls bathed in a warm glow that filters through the tall windows. The air is thick with the tantalizing aroma of exotic spices and fresh herbs, each scent intermingling to create an olfactory symphony that mirrors the cosmic patterns swirling in Jaxon's mind. He stands at the kitchen island, surrounded by an array of ingredients that are a testament to both his culinary past and his recent transcendental revelations.

Jaxon clutches a sketchpad, the pages covered in intricate diagrams of fractal patterns and culinary concepts. Each sketch represents a dish, meticulously designed not just to please the palate but to convey the very essence of the universe's interconnectedness. He moves a pencil across the paper with a steady hand, guided by a newfound clarity that the recent events have bestowed upon him. As he draws, he envisions how the geometric forms can be translated onto a plate, each dish a visual and sensory manifestation of the cosmic order he now understands so deeply.

His mind flashes back to the lightning strike, a moment that has been seared into his consciousness like a brand. It had realigned him, not just neurologically but spiritually, opening gateways to dimensions where matter and meaning meld seamlessly. This awareness fuels his current endeavor, as he believes that food can be a medium to share these insights, bridging the gap between the material and the metaphysical.

Jaxon sets down the sketchpad and begins pulling ingredients from the shelves with a sense of purpose and joy that he hasn't felt in years. His kitchen, once a place of mundane routine, now hums with the energy of a workshop where art and science fuse. He arranges vibrant reds of tomatoes, the lush greens of kale, and the deep purples of eggplants, creating a visual feast that hints at the deeper truths each component holds.

With the precision of an alchemist, Jaxon experiments with ingredients. He slices, chops, and mixes, each motion fluid and deliberate, reflecting the fractal symmetry that guides his every action. The sizzle of onions hitting the hot pan, the rhythmic chop of herbs, and the gentle clink of utensils together form a melody that resonates with the patterns of the cosmos. Jaxon allows his movements to flow naturally, entrusting the process to his enhanced intuition.

He prepares a dish that aligns the principles of sacred geometry and quantum physics, plating it with carefully placed layers that mirror the intricate designs of his fractal sketches. The papery skin of an onion reveals successive layers, a microcosm reflecting the larger structures of reality he has come to understand. Twirls of pasta echo the spirals of galaxies, while sauces are applied in cascading flows that mimic the ebb and flow of cosmic currents.

As each dish takes form, Jaxon feels a profound sense of connection to the universe. The kitchen, with its scents and sounds, becomes a conduit for his newfound purpose. His heart beats in synchrony with the rhythmic chopping and stirring, his breath aligning with the simmering fusion of flavors. Each taste, a journey through the labyrinth of his experiences, holds the promise of conveying something transcendent to those who partake.

Once the dishes are ready, Jaxon sets about inviting those who seek wisdom to a gathering filled with hope and purpose. He sends out messages, written with careful consideration, to individuals who resonate on the same frequency of curiosity and introspection. Each invitation is a call to experience an exploration of consciousness through culinary art, an intimate offering of the cosmic truths Jaxon has uncovered.

With a soft smile, he reflects on the journey that has led him to this point. His fall from grace, the grasping clutches of addiction, and the bolt of lightning that rewired his very essence–all these moments converge in this kitchen. The once sorrowful lament of his past now echoes as a harmonious note within the symphony of his creation. He understands that food, an elemental need, can transcend to become a spiritual tool, guiding others toward enlightenment and connection.

As the morning sunlight floods the kitchen, Jaxon surveys the array of vibrant dishes laid out before him. Each one is a testament to his journey, a tangible expression of the fractal patterns that govern both the universe and the human soul. The colors and shapes resonate with life, their aromas mingling to create an atmosphere charged with the energy of new beginnings.

Jaxon's heart swells with anticipation and a sense of fulfillment. The textures, flavors, and visual beauty of his culinary creations are not just a feast for the senses but an invitation to experience the interconnectedness of all things. Through this gathering, he hopes to share a taste of the infinite, allowing others to glimpse the deeper layers of reality and their place within it.

As he stands in his kitchen, a place that has witnessed both his darkest moments and his most profound revelations, Jaxon feels a renewed sense of purpose. The new chapter that dawns with the morning light is one of harmony and integration, where his culinary art becomes a beacon of spiritual insight and communal connection. Smiling at the prepared dishes, he knows that this is just the beginning–his journey of sharing wisdom through food is a path that extends endlessly, like the fractals that inspire it.

Chapter 16

A CONSTELLATION OF FLAVORS

    Jaxon stands in his newly established kitchen, a world infused with promise, surveying the myriad of spices and tools that line the cluttered shelves. The space is bathed in warm, ambient light, casting soft shadows that give the room an almost mystical atmosphere. He breathes in deeply, the mingled scents of cardamom, cumin, and rosemary awakening dormant memories of his previous life, one filled with brilliance and unbridled creativity. Each breath fills him with a sense of renewal, the air itself charged with the possibilities that lie ahead.

The kitchen is his sanctuary, a place where he can transform the chaos of his mind into something tangible, something beautiful. As his fingers trail over the containers, he contemplates how each ingredient holds the potential to create dimensions within a dish. He marvels at the complexity of nature, how a single spice can carry the essence of an entire landscape, how a herb can encapsulate the soul of a region. The kitchen, once a battlefield of stress and deadlines, has now become his temple—a sacred space where he can explore the depths of his soul and the universe.

Pausing, Jaxon selects a ripe tomato from a basket, the red skin smooth and reassuring under his touch. He closes his eyes momentarily, feeling the textures and visualizing the vibrant patterns of nature's symmetry. The hues and rhythms of the produce echo the fractal insights he has unlocked; the connections between the folds of existence now seem as tangible as the vegetables in his hands. Each cut, each selection, is deliberate, a dance with the cosmos.

Methodically, Jaxon clears the counter, aligning various knives, cutting boards, and bowls with precision that mirrors the clarity he now feels within. He wipes the surface down, every stroke a cleansing ritual, transforming chaos into order. The haphazard clutter that once dominated his life has no place here; his workstation is an extension of his mind, immaculate and brimming with possibilities. As he arranges the mise en place, he channels his emotions into this labor, seeking to manifest his revived spirit into a physical form.

The kitchen hums with energy, the low sound of simmering pots and the rhythmic chop of knives creating a soothing melody that mirrors the natural rhythms of the universe. Jaxon's thoughts flow effortlessly as he works, his hands moving with a confidence that had once been lost to him. The act of cooking becomes a meditation, a way to align his mind, body, and spirit with the greater forces at play in the cosmos.

Settling into the rhythm, Jaxon sets a pot on the stove and watches as water begins to boil. The bubbling liquid reminds him of the effervescent flow of thoughts now coursing through his reshaped consciousness. He dices onions with a practiced grace, the knife movements fluid, almost meditative. Each slice feels harmonious, as though the universe speaks through the rhythmic cadence of his culinary choreography. His senses heighten–savory scents mingle with the soft sizzle of ingredients, creating an auditory symphony in the kitchen.

As the aromas intensify, Jaxon's mind drifts to his estranged sister, Angelica, and the pain of their fractured relationship. The guilt over past behaviors gnaws at him, driving him to find redemption through his craft. He envisions them sitting at a table, sharing a dish that transcends mere nourishment, offering a communion of souls seeking reconciliation. The dish he prepares is not just food; it is an olive branch, an offering of peace and understanding. Every ingredient, every technique, is infused with the desire to heal the wounds that have kept them apart.

The anticipation of reconnecting with Angelica fuels his creativity, pushing him to perfect every element, to craft something imbued with both taste and meaning. He imagines the look on her face when she takes the first bite, how the flavors will transport her back to their childhood, to a time before their lives diverged into separate paths. The kitchen becomes a portal to the past, a place where he can reclaim the connections he has lost.

As he continues to cook, the kitchen fills with an intoxicating aroma, each scent layered like the notes of a complex song. Olive oil dances with crushed garlic in the pan, releasing a pungent fragrance that blends with the earthy sweetness of roasting peppers. He sprinkles spices with reverence, allowing their essences to weave intricate patterns that echo his cosmic revelations. The act of cooking becomes a spiritual practice, a meditation on unity and transformation.

Time seems to dilate as Jaxon moves seamlessly from one task to another, lost in the joy of creation. He stirs a rich, velvety sauce, the wooden spoon tracing curves that resemble the fractal geometry occupying his thoughts. The ingredients meld into a harmony of flavors, a testament to the interconnectedness of all things. His heart races, not with anxiety, but with the exhilaration of discovery, each moment in the kitchen a step further into a new realm of culinary artistry.

He places the final touches on the dish, arranging herbs and edible flowers with an artistic flair. A sense of peace washes over him as he surveys his creation, the culmination of his journey from addiction to enlightenment. The colors burst forth vividly, each component a brushstroke on the canvas of his culinary masterpiece. The dish is more than food; it is a narrative, a reflection of his reconciliation with the universe and himself.

Stepping back, Jaxon feels a profound sense of accomplishment. The anticipation of sharing this creation with others, of translating his inner growth into a sensory experience, fills him with hope. He looks around his kitchen, now a sanctuary of inspiration and clarity, and for the first time in years, feels the pulse of excitement for what lies ahead. The journey has just begun.

In the early evening, Jaxon stands in the heart of his kitchen, a sanctuary of innovation and harmony. The light outside is soft, casting a golden glow through the windows, as if the day itself is winding down in anticipation of what is to come. His fingers dance across various playlists, experimenting with genres that range from classical symphonies to modern electronic beats. The kitchen hums with life, the stainless steel appliances reflecting the golden hues of the pendant lights above, casting a warm glow across the room.

As orchestral crescendos fill the space, Jaxon feels the ancient roots of culinary traditions melding with the present. The resonant sounds of a violin solo transport him to the grand feasts of Renaissance courts, where opulent banquets were a testament to both artistic expression and communal pride. He recalls how these historical feasts celebrated the convergence of flavor and form, each dish a masterpiece meant to evoke not just taste, but a sense of belonging and wonder.

The music shifts to rhythmic African drum beats, and he senses the pulse of tribal gatherings, where food was a conduit of community and spirituality. The deep, earthy tones ignite visions of communal fires and shared stories, a reminder of food's role in binding human connections. The power of rhythm inspires him to select ingredients for his dish–technicolor peppers, heirloom tomatoes, and verdant herbs. He contemplates the frequencies of their colors, imagining the sound each one would produce in harmony with the others.

Jaxon's mind travels through these epochs, each genre unlocking a different facet of culinary history, from the minimalist elegance of Japanese Kaiseki to the robust, aromatic flavors of Indian thalis. These culinary philosophies, though born in disparate corners of the world, shared a common thread: the belief that food could transcend its physical form to touch the soul. The ambient music now playing–soft, ethereal tones–reflects his desire to infuse these ancient principles with his newfound cosmic insights.

He arranges the vibrant produce on the countertop, considering how each ingredient's natural resonance will shape the dish. Celery stalks become melancholic cellos, their crisp green rhythms softened by the deep octaves of earthy beets. Sweet corn kernels, like sparkling piano notes, brighten the composition. His knife glides through each ingredient with precision, the rhythmic chop and slice becoming a dance of creation. The sound of a chef's knife echoing against the wooden cutting board feels hypnotic, a mantra of culinary alchemy.

Jaxon reflects on his journey, where addiction once dulled his senses and clouded his mind. The cosmic realignment he experienced after the lightning strike has not just awakened his culinary genius; it has reshaped his very understanding of existence. He ponders how sound and flavor, intertwined, can elevate not only the gustatory experience but also serve as a medium for expressing deeper cosmic truths.

As he garnishes the plate, Jaxon hums softly, a melody that seems to vibrate through his soul and into the dish itself. He recalls family gatherings from his childhood, the music his mother would play blending seamlessly with the aromas of traditional recipes. Each note he hums evokes memories of laughter around the dinner table, the reassuring warmth of those familial bonds intermingling with the uncertainty and creativity of his present.

Finishing the dish, Jaxon takes a step back, admiring how the colors glow with an almost otherworldly harmony. He views his culinary creations as fractal expressions of the universe–intricate, multifaceted, and infinitely connected. The plate before him mirrors this complexity; it is not just a meal but a homage to the cosmic web that binds all existence together. The dish seems to pulse with life, colors blending in a harmony that speaks to the very essence of the cosmos.

The kitchen, once a place of mere sustenance, has become a sacred space where Jaxon can explore the deeper truths of existence. Every ingredient, every tool, is imbued with meaning, each one playing a role in the grand symphony of his creations. The air is thick with the scent of herbs and spices, a sensory reminder of the journey he has undertaken.

Setting the completed dish aside, he feels an overwhelming sense of connection to the energy that flows through everything–the same energy that orchestrates symphonies and binds the elements into a coherent whole. His kitchen, filled with the sensory richness of sizzling pans and melodic hums, becomes a microcosm of the vast universe. Jaxon contemplates his next steps, eager to continue this journey where culinary art becomes a sacred bridge to the cosmos. As he gazes at the vibrant presentation, he recognizes that his culinary creations are more than sustenance–they are a glimpse into the infinite, a taste of the divine.

Jaxon sits at his polished kitchen table, surrounded by a fortress of books and digital screens. The soft glow of the pendant lights above casts a golden hue over the pages he voraciously reads, and the digital sketches he meticulously draws. His fingers dance across the smooth surface of the table, tracing invisible patterns. Fractals spin in his mind, geometric shapes folding and expanding, each one a gateway to the mathematical beauty he seeks to bring to life through his cooking.

He leans back, taking a moment to stretch, his eyes wandering around the transformed kitchen. Every shelf, every corner, resonates with his newfound clarity. It's more than a kitchen–it's an art studio, a laboratory, and a sanctuary all rolled into one. Aromas from earlier experiments linger in the air–citrus zest, fresh basil, a hint of cardamom–each scent a testament to his relentless quest for harmony.

Jaxon's hand hovers over a selection of plates, each one uniquely crafted to highlight the fractals he envisions. He takes a deep breath, feeling the smooth ceramic beneath his fingers, and begins to plate his food. The process feels like nature's unveiling–careful layers of vegetables, splashes of color from sauces, meticulous arrangements of herbs. He moves with a fluidity that mirrors the fractals in his mind, each motion a purposeful dance of creation.

With the first dish complete, Jaxon steps back to admire his work. The harmonious blend of textures and colors mirrors the complexity of the cosmos, a visual symphony that hints at the flavors waiting to be experienced. He feels a surge of satisfaction but knows the real test is yet to come. He invites his friends to join him, their presence a necessary ingredient in this multifaceted experiment.

The dining area hums with anticipation as his friends trickle in. The table, set with elegant simplicity, glows under the soft light, ready to host an evening of culinary exploration. Jaxon's heart swells as he observes Maya, Dr. Miles, Allegra, and Angelica taking their seats. Each person represents a different part of his journey–Maya's philosophical inquiries, Miles's scientific skepticism, Allegra's mystical aspirations, and Angelica's grounding love. Their interactions create a rich tapestry of connection and insight.

Jaxon begins to serve his friends, each dish an infusion of his cosmic revelations. As the plates land on the table, he watches their reactions with keen interest. Maya gazes at the intricate designs, her eyes wide with wonder. Dr. Miles scrutinizes the fractal patterns with analytical curiosity, his

mind already formulating questions. Allegra takes a deep breath, her fingers brushing the edge of the plate as if feeling the vibrations of the universe within. Angelica, ever supportive, beams at Jaxon, her eyes shimmering with pride.

The first bite is taken in silence, the room thick with expectation. Jaxon holds his breath, his muscles tense as he reads their faces. Maya is the first to speak, her voice filled with awe. "This is... more than food, Jaxon. It's a journey."

The conversations flow, their voices filling the room with a mix of excitement and contemplation. Dr. Miles leans in, his skepticism softened. "The balance of flavors is remarkable. It's like tasting an equation."

Allegra, usually reserved, allows herself a rare smile. "You've managed to capture the essence of the divine in every element. It's transformative."

Angelica's feedback is simple but heartfelt. "I feel your journey in every bite. It's beautiful, Jaxon."

As he cleans up, Jaxon listens to their reflections, feeling a profound sense of fulfillment. Each comment, each expression, is a piece of the mosaic he's been crafting. His friends' feedback is more than validation–it's a mirror reflecting his success in merging his culinary skills with his spiritual insights. The dishes are not just food; they are stories, each one a chapter in his narrative of transformation and redemption.

The evening winds down, and Jaxon watches his friends' animated discussions with a quiet joy. Their reactions, their understanding of his deeper message, weave a sense of unity and purpose. It's in this moment that Jaxon truly grasps the impact of his work. His culinary creations have transcended the physical, touching the minds and souls of those who partake.

As the night deepens, Jaxon stands back, a surge of gratitude washing over him. The fractal journey on his plates symbolizes the interconnectedness of everything–every flavor, every texture, every moment binding the past, present, and future into a coherent whole. His friends leave, their hearts and minds enriched by the experience, and Jaxon is left in the serenity of his kitchen.

Jaxon gazes at the aftermath of the divine feast–the empty plates, the lingering fragrances, the echoes of laughter and deep conversation. He knows this evening marks a new beginning. He feels a renewed sense of purpose, a forward momentum that propels him into the next chapter of his journey, ready to share his cosmic culinary art with the world.

Jaxon stands before the mirror, the soft glow of the kitchen light illuminating the lines and contours of his face. He runs a hand through his tousled hair, now flecked with silver strands that hint at both age and wisdom. His green eyes, once dulled by the fog of addiction, now sparkle with clarity and reflection. The scar on his cheek, a stark reminder of a past life in the chaotic turmoil of kitchens, glistens faintly, a testament to his survival and resilience. He takes a deep breath, feeling the rise and fall of his chest, each inhalation a reminder of his journey from darkness to this moment of lucidity.

Before him lies his journal, open to a blank page, inviting the flow of thoughts and memories. He picks up the pen, feeling its familiar weight in his hand, and begins to write. The words come easily, each stroke of ink a nod to his experiences. He details how the fractal patterns he now sees in his mind's eye have woven themselves into his culinary creations, a fusion of geometric perfection and spiritual insight. These patterns, born of his deeper understanding of the universe, find their place in the delicate arrangement of ingredients, in the harmonious blend of flavors and textures. As the pen glides over the paper, he recalls the way music plays its role, how different notes and rhythms influence not just his mood, but the essence of each dish. His cooking has become an alchemical process, a symphony of cosmic connections manifesting in the culinary art.

He pauses, eyes drifting to the corner of his journal where sketches of fractal designs intertwine with musical notes. Each drawing, a mini cosmos of its own, speaks of an intricate, interconnected reality. His thoughts wander back to moments of hardship, the nights spent in the grip of addiction, and the subsequent lightning strike that realigned his consciousness. These memories, once sources of pain, are now embedded with gratitude. Jaxon understands that without the struggle, he would not appreciate the present clarity, nor the authenticity he has reclaimed.

He writes about his encounters with others, about the seekers who came to him, each carrying their own burdens and aspirations. He details how his interactions with them have shaped his journey, how their questions and reflections have pushed him to delve deeper into the mysteries of the universe. He acknowledges the role they have played in his transformation, how their presence has been both a challenge and a blessing.

Closing his journal, he gazes out of the window. The view beyond is serene, the night sky a canvas of scattered stars and a glowing moon, each celestial body whispering tales of the universe. The quiet hum of the city below contrasts the expansive stillness above, a duality that mirrors his own journey. As he breathes in the crisp night air, he sees the world not just as it is, but as a mosaic of possibilities, a vast canvas for his culinary explorations. The city lights twinkle like the sparks of innovation, each one a beacon guiding his path.

Jaxon's kitchen, his sanctuary, is now a vibrant space filled with colors, textures, and aromas that resonate with life. Shelves brimming with spices and fresh herbs exude a symphony of scents, while the polished counters hold the traces of his latest creations. Every utensil, every ingredient, serves as a tool for his artistry, each one holding the potential to transcend the ordinary. The gentle light bouncing off the marble surfaces creates an atmosphere of warmth and creativity, enveloping him in a cocoon of inspiration.

He moves to the kitchen, feeling the coolness of the tiles beneath his feet, and begins to prepare a simple dish. The act of cooking, once a means to an end, has become a meditative practice, a way to connect with the deeper rhythms of life. As he chops vegetables and stirs sauces, he reflects on the journey that has brought him here. The kitchen, once a place of stress and exhaustion, is now a sanctuary of creativity and peace.

Closing his journal, he steps back from the window and smiles, feeling a profound sense of peace and fulfillment. The night sky mirrors his inner landscape, a vast expanse of possibilities and newfound purpose. His culinary studio stands ready, a testament to his journey and the knowledge he has gained. In this moment, Jaxon embraces both his past and future, understanding that every experience, every hardship, has led him to this point of harmony.

As he gazes at the stars, he feels a deep connection to the cosmos, an unspoken promise of continued exploration and growth. The universe stretches out before him, an endless source of inspiration for the dishes he will create, each one a bridge between the material and the spiritual. With a renewed sense of purpose and excitement, Jaxon closes his eyes, letting the silence of the night seep into his being, knowing that his journey, both as a chef and as a seeker of truth, has only just begun.

Chapter 17

THE LUMINOUS PATH

Jaxon steps into his culinary studio, the space bathed in golden hues filtering through the wide-paned windows that overlook the bustling city below. The studio, once a vacant loft, has been meticulously transformed into a sanctuary of culinary exploration. Every surface gleams with purpose, from the polished countertops adorned with fresh ingredients—vivid red tomatoes, fragrant basil leaves, bulbs of garlic, and an array of spices in every conceivable shade—to the carefully curated utensils and tools, each with a specific role in the alchemy of cooking.

The air is thick with the aroma of freshly baked bread mingling with the subtler scents of herbs and spices, creating an inviting tapestry for the senses. Soft, ambient music hums in the background, its gentle melodies crafted to soothe the soul and calm the mind. Each note resonates with the intention behind this space—a place where the culinary and the cosmic converge, where food becomes more than sustenance; it becomes a metaphor for transformation and healing.

Jaxon moves through the room with a quiet confidence, arranging the last of the jars and bottles, his hands moving with the practiced grace of someone who has spent a lifetime in kitchens. But this space is different. It's not just about creating dishes; it's about creating experiences, guiding others through a journey of self-discovery much like the one he has endured. His thoughts drift to the purpose behind this endeavor–a redemptive arc, a path through which he hopes to navigate not just his own resurrection but to guide others toward their own awakenings.

He pauses, looking around at the sanctuary he has created, and inhales deeply, the scent of basil and thyme grounding him in the present. This studio, with its warmth and light, is a far cry from the chaos of his past–the frenetic kitchens, the relentless pursuit of perfection that led him down a path of addiction. Now, he seeks to use this space to bring others closer to their own truths, to find peace in the act of creation.

The creak of the studio's front door pulls Jaxon from his reverie. The seekers begin to file in, each one bringing their own energy, their own story, to this sacred space. They are a diverse group, reflecting a tapestry of human experience. There's Maya, whose deep-set eyes reflect both skepticism and curiosity; Dr. Miles Carter, whose air of scholarly rigor carries a touch of apprehension; and Allegra Frost, whose ethereal presence is both commanding and intriguing. Each one of them steps into the studio, their expressions a blend of excitement and trepidation, the air ripe with the possibilities of the day ahead.

"Welcome," Jaxon says, his voice warm and inviting. He takes deliberate steps toward the seekers, extending a hand to each one, the simple act dissolving any lingering barriers. "I'm thrilled to have you here. Today, we embark on a journey, not just through flavors, but through self-discovery and transformation."

He gestures for them to take their places around the high marble island in the center of the room. "This space," he continues, "is a sanctuary. Here, food is not merely sustenance but a pathway–a medium through which we explore the depths of our own consciousness and begin to heal."

The seekers exchange glances, some nodding, others murmuring in agreement. Maya, always the philosopher, is the first to speak. "I'm here because I've always believed in the transformative power of experience. Food, art, spirituality–they're interwoven, and I want to see how deep that connection goes."

Next to her, Dr. Miles Carter adjusts his wire-rimmed glasses. "I've approached this with a degree of skepticism, I admit," he says. "But I'm also deeply curious. Your equation and its implications have raised questions I can't ignore."

Allegra stands poised, her eyes sparkling with fervor. "For me, this is not just about learning; it's about ascending to higher planes of awareness. Jaxon, your journey is a testament to resilience, and I'm here to witness and participate in that transformation."

Jaxon listens, his heart swelling with gratitude. These are not just participants–they are allies, each bringing their unique perspectives to the table. He reflects silently on how different each seeker's journey is from his own, yet how unified they are by the shared desire for growth and understanding.

"Thank you for your honesty," Jaxon replies, his tone sincere. "Today, we will cook together, but more importantly, we will share with each other. I want you to think of each ingredient not just as a component of a dish but as a metaphor for an aspect of your life. The process, the intention, the mindfulness you bring into this kitchen will mirror your personal journey."

He pauses, taking in their receptive faces before continuing. "Let's start by sharing a bit more about ourselves. Why are you here? What are you seeking to discover or overcome?" His open-ended question hangs in the air, inviting each seeker to delve deeper into their motivations.

The group begins to open up, stories spilling forth of struggle, aspiration, loss, and hope. As each person shares, Jaxon feels the weight of their words, the gravity of their experiences. He doesn't just listen; he absorbs, understanding that this moment is as vital to his journey as it is to theirs.

The studio hums with an undercurrent of anticipation. It's a mingling of stories and souls, created not just by the aroma of fresh ingredients but by the palpable sense of shared purpose hanging in the air. As Jaxon listens, he reflects on the weight of his past failures and the soaring hopes he places on this new venture. The stakes are high, not just for himself but for every individual who steps into this space seeking something more.

The atmosphere, scented with basil and thyme, is warm and welcoming, casting a soft glow on the faces of the seekers. The studio, once a mere physical location, now feels like a living entity, breathing with the collective aspirations of its inhabitants. The basil's sharp sweetness mingles with the earthiness of thyme, creating a sensory tapestry that speaks to warmth and possibilities. In this sacred space, amidst the clinks of knives and the whisper of ingredients being prepared, the seekers find not a chef, but a mentor, a guide who understands the intricate dance between flavor and feeling, between the culinary and the cosmic. As the day stretches out before them, filled with promise and potential, Jaxon stands ready, listening attentively, fostering a community bound by the desire for deeper understanding and transformation.

The culinary studio, bathed in the warm afternoon light filtering through tall windows, envelops the seekers as they gather around the central island. Fresh ingredients line the countertops, a vibrant mosaic of possibilities: vivid bell peppers, crimson pomegranates, aromatic herbs, and exotic spices. The air is thick with the scent of freshly chopped basil and rosemary, inviting them into a world where culinary artistry meets spiritual exploration. Jaxon stands at the helm, exuding a sense of calm authority, his eyes reflecting the depth of his journey.

"This dish," Jaxon announces, placing a beautifully plated creation before the group, "is called 'Ancestral Harmony.' Each ingredient holds a symbolic meaning, intertwined with my path to self-discovery." He gestures to the vibrant array on the plate–a delicately roasted beet lies at the center, its rich, earthy hues a testament to grounding and roots. Surrounding it, julienned carrots and slivers of golden apple form a spiral, symbolizing the fractal nature of growth and evolution. A drizzle of honey and balsamic reduction glistens like an alchemical elixir, merging sweetness with the sharpness of understanding.

The seekers lean in, mesmerized not just by the dish but by the narrative Jaxon weaves. Their expressions range from curiosity to contemplation, each one silently linking Jaxon's reflections to their own struggles and aspirations.

The soft glow of pendant lights adds a layer of intimacy, casting gentle shadows that dance across their faces.

Emboldened by his own story, Jaxon encourages the seekers to explore their creativity. They break into small groups, their excitement palpable as they gather around different stations. Conversations spark as they discuss ingredients and flavor profiles, their voices mingling with the rhythmic chopping of vegetables and the sizzle of olive oil hitting hot pans.

Maya, with her braid swinging as she moves, quickly takes charge of her group's station. "Let's create something that symbolizes transformation," she suggests, her eyes bright with enthusiasm. "Perhaps a salad with layers and textures that evolve with each bite?" Around her, the group nods, inspired by her vision.

Meanwhile, at another station, Dr. Miles Carter, his analytical mind now open to the experience, examines a variety of mushrooms. "The diversity of mushrooms could represent the interconnectedness of our experiences," he muses, arranging them with precision. His group listens intently, their own ideas blossoming under his guidance.

As the seekers immerse themselves in the alchemy of cooking, the studio hums with energy. Pans clang, knives slice through crisp produce, and the bubbling of simmering sauces fills the air. The scent is intoxicating—a melange of garlic, citrus, and hints of exotic spices, creating an atmosphere where every sense is engaged.

Jaxon moves from group to group, his presence reassuring and inspiring. "Remember," he says gently to a young woman meticulously zesting an orange, "it's not just about the taste. Each movement, each decision you make, carries intention. Let your dish speak your journey." She smiles, her previous nervousness melting away as she layers the zest onto her creation with newfound confidence.

In another corner, a soft laughter erupts as two seekers accidentally spill a bowl of quinoa. Instead of frustration, they share a conspiratorial grin and energetically begin again, their camaraderie growing stronger. Each group, despite their individual challenges, finds unity through their shared efforts.

The transformations each seeker seeks begin to surface in their conversations. One man speaks of his struggles with self-worth, another woman describes her recent journey of healing from loss. The act of cooking together becomes a medium through which they articulate their innermost desires and challenges, each ingredient a piece of their evolving narratives.

The sounds of the studio transform into a symphony of creation, underpinned by the occasional burst of laughter or murmured word of encouragement. The warmth of the room, both literal and figurative, holds them in a comforting embrace. The studio has evolved into a sanctuary–a place not merely of culinary exploration but of personal revelation and connection.

Pots across the studio bubble with exuberance, creating a harmony that mirrors the emotional boiling and blending occurring among the seekers. Jaxon, the vigilant conductor of this symphony, guides his seekers not just with expertise but with empathy, illuminating their path with the wisdom gleaned from his own tumultuous journey.

As the melodies of culinary creation reach their crescendo, Jaxon stands back and takes in the scene. The once disparate individuals now move with an almost choreographed grace, each action infused with intention and mindfulness.

The studio has become a lively hub of activity and conversation, pots bubbling and laughter echoing. The seekers, united by the common purpose of culinary and spiritual exploration, find solace and strength in the bonds they forge, their personal transformations simmering to the surface with every savory creation.

The soft clinking of utensils and murmur of conversation drift through the room as Jaxon surveys the studio. There's a glow in the air, a mingling of candlelight and the fading day that casts whimsical shadows on the walls. The aroma of simmering broths and freshly chopped herbs wafts through the space, creating an inviting warmth that beckons everyone to pause and breathe deeply.

He gathers the group in a semicircle, chairs pulled close together, evoking the intimacy of old friends sharing secrets around a campfire. Jaxon's voice is calm yet vibrant, drawing everyone's attention. "Let's take a moment to reflect on our experiences today," he begins, his gaze moving from face to face, noting the variety of backgrounds and stories etched into each expression. This sea of diversity, woven together by shared aspiration, stirs a sense of purpose within him.

"Think about what you've learned while cooking. How did it feel? What did it reveal about your journey?" His words float through the air, inviting vulnerability, encouraging the seekers to delve deep. Jaxon's own journey lingers at the edges of his mind–a path marked by addiction and self-discovery, each phase a crucial stitch in the fabric of his existence.

One by one, the seekers begin to share. A woman with cropped, silver-streaked hair talks about the meticulous slicing of vegetables symbolizing her need for control and order in a chaotic life. A young man speaks reverently of the alchemy he felt while blending spices, a metaphor for his desire to integrate disparate parts of his fractured self. Jaxon listens intently, recognizing pieces of his own story in their reflections, the raw honesty of their revelations mirroring his internal struggles.

Jaxon then introduces the concept of mindfulness, drawing parallels to ancient spiritual practices. He emphasizes how, in a world overflowing with distractions, the act of being present is a rebellious act of self-love. "Mindfulness in cooking isn't just about the food," he says. "It's about connecting each action with an intention, creating a dish that not only nourishes the body but also the soul."

As he speaks, memories flash through his mind–the chaotic tumult of kitchen life, the relentless pursuit of perfection that often led him astray. It was only through meditative practices that he found solace, a bridge that united his past with his present understanding. Meditation offered glimpses of clarity that he now wishes to share, hoping to see that same light dawn on the faces around him.

Jaxon leads the seekers in a brief meditation, their breaths synchronizing with the rhythm of the universe. Eyes closed, they inhale the mixed scents of herbs, garlic, and simmering sauces, allowing these aromas to anchor them in the moment. The gurgling of a nearby fountain adds a soothing backdrop, blending seamlessly with the soft hum of the distant city.

The room falls into a profound silence, each person attuning to their inner world, unearthing intentions buried under layers of daily noise. Here, in the calm, they rediscover fragments of themselves, integrate their cooking experiences, and find new connections tied to their personal truth. As the meditation ends, Jaxon opens his eyes and meets the gaze of each seeker. "Share what insights you've discovered," he encourages gently. Hands slowly rise, voices finding their way back into the room, each person articulating the transformations they've begun to experience. There's a newfound softness in their tones, a communal understanding that has blossomed from their shared silence.

The seekers' stories weave an intricate tapestry of human experience, each thread adding depth to the collective understanding. A woman speaks of how kneading dough reminded her of resilience–each press and fold a metaphor for bending but not breaking under life's pressures. Another shares how dicing onions led to tears, both from the vegetable and from releasing long-suppressed emotions.

Jaxon feels a surge of gratitude as he watches the group, their individual transformations reflected in the dishes they've prepared. Each creation is a testament to their journey, a tangible manifestation of their inner growth. He senses the energy in the room shifting, evolving from cautious curiosity to palpable camaraderie.

The studio, bathed in the waning light of evening, resonates with unity and understanding. The flickering candles cast dancing shadows on the walls, and the scent of their culinary creations fills the air with a sense of accomplishment and promise. Jaxon's heart swells with a sense of fulfillment. This is precisely why he opened his doors–to foster an environment where the act of cooking becomes a pathway to deeper self-awareness.

"Remember," Jaxon concludes, his voice imbued with quiet authority, "this journey doesn't end here. Take these insights with you. Let them guide you, not just in the kitchen, but in every aspect of your life."

The seekers nod, absorbing his words, a mutual understanding passing among them. As the session draws to a close, the room buzzes with the warmth of shared purpose, each person ready to carry forward the lessons learned.

Jaxon stands back, observing the connections formed and the growth initiated. Amidst the hum of voices and clinking dishes, he feels an unspoken assurance—a knowledge that the day's journey is just the beginning of something profound.

Jaxon watches as the seekers arrange the communal dining area, transforming the studio into a vibrant, inviting space. The long wooden table is draped with an array of colorful linens and adorned with small, flickering candles. Warm lighting from the pendant lamps casts a golden hue over the room, mixing with the amber glow of the setting sun streaming through the windows. The inviting aroma of freshly prepared dishes fills the air, creating a tapestry of scents that evoke both the comfort of home and the excitement of exploration.

Each seeker takes a turn presenting their dish with a mix of pride and nervous anticipation. The table brims with a diverse array of culinary creations, each offering a glimpse into the cultural backgrounds and personal journeys of their creators. Natalia, a young woman from Brazil, stands first, explaining how her vegan feijoada—a traditional black bean stew—represents her commitment to sustainability and her newfound connection to the earth. Her eyes shine
as she talks about the pride she feels in embracing her heritage in a way that aligns with her values.

Next, Ahmed, a software engineer from Egypt, presents koshari, a hearty mix of lentils, rice, and pasta topped with a tangy tomato sauce. He shares how the process of making this dish has mirrored his path to reconciling his dual identities—one rooted in ancient traditions, the other navigating the rapid pace of modern technology. His voice wavers slightly with emotion as he describes how cooking has helped him to find a sense of unity within himself.

As each seeker presents their dish, the atmosphere becomes charged with shared stories and personal revelations. The clatter of utensils and the hum of anticipation create a symphony of activity, punctuated by laughter and the occasional clink of glasses. Sophie's French ratatouille is not just a dish but an homage to her late grandmother, who taught her that true artistry lies in simplicity and care. She speaks softly, her voice a melody of nostalgia and gratitude, evoking a gentle round of applause from the group.

Jaxon circulates among the seekers, his presence a blend of mentor and fellow traveler. He offers words of encouragement and insightful questions, drawing out the deeper connections between their culinary choices and their inner transformations. His gratitude is palpable, a warm current that flows through the room, subtly knitting the group closer together.

As the final rays of sunlight give way to dusk, the group settles down to savor the feast. The dining area, now buzzing with lively conversation, resembles a microcosm of interconnected lives and shared experiences. They discuss not just the flavors and textures of their dishes but the profound self-discoveries they have made through this culinary journey. A sense of camaraderie strengthens their bonds, laughter mixing with reflective pauses.

Jaxon raises his glass, capturing the attention of the room. "I want to thank each of you," he begins, his voice filled with sincerity. "For your openness, your courage, and for sharing pieces of your soul through your creations. This studio is not just about food–it's about connection, growth, and the journey we all embark on."

There is a murmur of appreciation and agreement as he speaks. He gazes around the table, seeing not just faces but stories of transformation, brief friendships that feel timeless. The sky outside shifts into deeper shades of purple and indigo, the candles casting dancing shadows that mimic the vibrant internal lives of the seekers.

As the evening progresses, subtle nuances of interpersonal relationships emerge. Ahmed and Sophie bond over a shared love of classical literature, while Natalia offers to teach a yoga session to anyone interested. There are moments of vulnerability, as when Maya opens up about her fear of never being good enough, prompting an outpouring of support and shared stories of overcoming self-doubt.

The meal concludes with a collective sense of fulfillment and a silent acknowledgment of the path that lies ahead for each of them. Jaxon's parting words linger in the air like the scent of spices, infusing the group with hope and a renewed sense of purpose. "This is just the beginning," he says, his eyes reflecting the dim candlelight. "Carry these lessons with you, and let them shape your journey."

As they leave the studio, the seekers take with them more than just the taste of meticulously crafted dishes; they depart with a piece of the infinite–an inkling of the fractal nature of their own consciousness. The studio, now quiet and dim, holds the echoes of their laughter and the scattered remnants of their shared feast, a sacred space where new realities were seeded.

Chapter 18

CONVERSATIONS WITH CHAOS

Jaxon steps into the culinary studio early, the morning light streaming through the expansive windows, casting a golden hue over the rustic decor of the space. The warmth of the wooden beams above and the soft tones of the walls create a sanctuary that contrasts sharply with the urban chaos outside. He breathes in deeply, savoring the aroma of fresh herbs and spices that linger from previous experiments.

The studio is meticulously organized, a living representation of Jaxon's journey from the darkness of addiction to a place of clarity and purpose. Every element, from the carefully placed pots and pans to the jars of vibrant spices, speaks of intentionality–a sharp contrast to the chaotic kitchens of his past. Each morning, Jaxon begins with a ritual: setting up the stations, selecting ingredients, and grounding himself through the methodical preparation of the day's menu. These small acts, performed in silence, have become his form of meditation, a way to align his spirit with the task ahead.

His hands glide over the array of ingredients laid out before him– vibrant vegetables, aromatic spices, and fresh herbs. The act of preparing these elements, one by one, speaks to the balance he seeks in his life, blending tradition with a newfound innovative spirit. Each knife stroke, each carefully measured spoonful of spice, is an exercise in mindfulness, a way to connect with the present moment and honor the process of creation.

While arranging a delicate line of saffron threads, the door swings open, and Maya bursts in, her presence electric and vivacious. The energy she brings with her is a sharp contrast to the calm atmosphere Jaxon had cultivated, yet it doesn't feel disruptive—instead, it adds a new dynamic to the space. Maya's entrance is like a gust of fresh air, revitalizing the studio with her enthusiasm and zest for life.

She surveys the scene, her eyes gleaming with mischief. "Early bird catches the worm, huh? Or in this case, the saffron threads," she teases, sliding into the kitchen with a playful energy that disrupts the serene stillness Jaxon has created.

Jaxon glances up from his work, a smile tugging at his lips as he continues to meticulously prep the ingredients, not missing a beat. "Someone has to keep it organized," he replies, his tone light yet laced with a quiet pride in the order he's maintained.

Maya grabs a pile of basil leaves and starts tearing them into a bowl, her movements fluid and spontaneous. "You know, Jaxon, sometimes the best flavors come from a little chaos. You can't just control everything."

Jaxon raises an eyebrow, still focused on his task but intrigued by her approach. "Is that so? Care to prove it? The challenge in his voice is light-hearted but genuine, his curiosity piqued by her confident spontaneity.

Their conversation quickly evolves into a spirited debate about the merits of structured techniques versus improvisation. Maya passionately advocates for the beauty of unexpected combinations, while Jaxon counters with the precision of classical methods. This exchange isn't just about cooking—it's a philosophical discussion that reflects their differing worldviews. Maya's belief in the power of spontaneity and Jaxon's commitment to meticulous craftsmanship highlight the tension between freedom and control, creativity and discipline.

"Cooking is like alchemy," Maya argues. "It's about intuition, feeling the energy of the ingredients and how they want to come together."

"True," Jaxon concedes, "but like any good alchemist, you need a solid foundation. Otherwise, you risk creating a mess instead of a masterpiece."

Their banter flows seamlessly into the act of cooking, the air around them charged with a vibrant synergy. As they begin to prepare a dish together, their hands move in unison, mixing Jaxon's disciplined methods with Maya's imaginative twists. The ingredients come alive under their touch, each motion a dance that conveys both their histories and their aspirations.

Tomatoes are diced into precise cubes by Jaxon, each slice deliberate and controlled, as though he's carving a piece of himself into each cut. Maya, in contrast, crushes a handful of garlic with a quick, decisive chop, her actions a testament to her impulsive nature, embracing the unexpected with each movement. The two styles meld, the combination yielding unexpected and delightful flavor profiles—a testament to the power of their collaboration.

As the rich aroma of simmering vegetables fills the air, they pause, glancing at each other. The exchange of looks speaks volumes, a mutual appreciation for the unique creativity they bring out in each other. Here in this studio, a space defined by natural light and warm, rustic elements, they find not just a place to cook but a sanctuary for growth and exploration.

The culinary studio represents a fusion of tradition and innovation, much like the evolving nature of their relationship. The blending of classical techniques with contemporary approaches mirrors their journey towards self-discovery and mutual respect. As they whisk sauces and season dishes, the act of cooking becomes a symbolic tapestry of their shared vulnerabilities and dreams.

In those glances and shared moments, Jaxon and Maya acknowledge the synergy they've created—a testament to the powerful intersection of their paths. It's a silent promise of what's to come, a contemplation of the infinite possibilities that lie ahead as they continue this journey together, both in the kitchen and beyond.

The sun climbs higher in the sky, its rays spilling through the studio windows and casting a warm glow on the polished countertops. Maya's knife moves swiftly, slicing through vibrant bell peppers. The colors—red, green, yellow—spill across the cutting board like paint on a canvas, each piece falling into its place with the precision of years of practice. Her movements are fluid, almost dance-like, reflecting years of ingrained family traditions and expectations.

A soft sigh escapes her lips as she begins to speak, the rhythmic chopping becoming a meditative backdrop to her words. "Growing up, my family had high hopes for me," Maya says, her voice a blend of vulnerability and determination. "My father, a philosopher, believed I would follow in his footsteps, while my mother, an artist, expected me to channel her creativity. Every decision felt like walking a tightrope, balancing their dreams and my own desires."

Jaxon listens intently, feeling the weight of her story in the air. The studio, bathed in warm, diffused light, seems to hold its breath. He senses the struggle within her, the silent battles fought within the walls of her home. He knows this path too well–the burden of expectations and the fear of failing those who matter most.

With each slice, Maya opens a window into her past. "There were days I wanted to escape, to find my own path. But their love was so strong, so overwhelming, it felt like a cage at times. I learned to blend my aspirations with theirs, like mixing ingredients until you can't tell where one flavor ends and another begins."

Her words resonate with Jaxon, who has experienced his own share of familial expectations and the struggle to find his true calling. He takes a breath, allowing her words to settle before sharing his own story. As he starts dicing onions, his mind wanders back to the haze of his addiction and the clouded brilliance that once defined him.

"I hit rock bottom more times than I can count," Jaxon begins, his voice steady but tinged with sorrow. "Addiction turned my life into a nightmare. I was a chef, a creator of experiences, but all I felt was emptiness. The kitchen, which once held my dreams, became my prison."

He pauses, the sharp scent of onions mingling with his memories. His hands move mechanically, each chop a tiny act of redemption. "I lost everything–my reputation, my friends, my family. My sister, Angelica, tried to help, but I pushed her away. I couldn't bear to see the disappointment in her eyes."

Maya stops slicing, her eyes locking onto his, filled with empathy and curiosity. He sees in her gaze a reflection of his own longing for connection, for understanding. The kitchen, with its rustic decor and the aroma of simmering possibilities, becomes their confession booth, their sanctuary.

"The lightning strike," he continues, "it wasn't just a physical shock. It realigned me, in ways I still can't fully grasp. The Divine Equation—the fractal paths it revealed—it's as much a journey through my own consciousness as it is a map of the universe." His voice quivers slightly, but he presses on. "It's taught me that every mistake, every moment of despair, was part of my transformation."

Jaxon's confession isn't just a recounting of his past—it's a declaration of his ongoing struggle to reconcile the man he was with the man he's becoming. His words are raw, unfiltered, and Maya listens with the same intensity with which she cooks, absorbing each syllable like a vital ingredient in a complex recipe.

As their conversation deepens, so does their cooking. Jaxon kneads dough with a focused intensity, each fold and press mirroring his internal struggle. The act of working with his hands, of feeling the dough transform under his touch, is therapeutic—an affirmation of his ability to shape his own future. Maya stirs a pot of simmering broth, the steam curling upwards, carrying whispers of their shared pain and hope. Their movements synchronize, an unspoken harmony that bridges their worlds.

Their collaborative dish begins to take shape—a canvas of textures and flavors reflecting their stories. The dough, infused with herbs, represents Jaxon's grounding process, returning to his roots while reaching for the cosmos. The broth, rich and aromatic, symbolizes Maya's journey, steeped in history and personal growth.

They taste as they go, making adjustments, adding spices and ingredients that speak to their emotional states. The dish evolves with each addition, becoming a living testament to their shared catharsis. When it's finally complete, the studio is filled with an intoxicating scent that blends their past and present, a fragrance that speaks of resilience, renewal, and the power of transformation. In silent understanding, they plate the dish, their faces reflecting a mix of pride and introspection. They take a moment of silence, their eyes meeting across the table. The weight of their experiences and the flavors they've created together hang in the air, a palpable connection forged in the crucible of their vulnerabilities.

The dish, savory and complex, tells their story through each bite. The crispness of the vegetables, the robustness of the broth, the comforting stability of the dough–all elements interplay, reflecting the intricate dance of their emotions. They taste it together, each bite a revelation, a recognition of their shared journey.

As they savor the dish in silence, a newfound depth settles between them. The studio, their sanctuary of creation and reflection, feels transformed, charged with the energy of their stories. The flavors linger on their tongues, a promise of what's possible when two paths intersect.

When they finish, Maya glances at Jaxon, a soft smile playing on her lips. He returns the look, grateful for this unexpected synergy. In this moment, they both understand that their connection is more than just a collaboration–it's a beginning, a canvas of infinite possibilities.

The silence between them is comfortable, a shared understanding that requires no words. They sit quietly for a while, letting the meal digest and the significance of the day's events sink in. The sun is lower in the sky now, casting long shadows across the studio floor. Outside, the city continues its relentless pace, but within these walls, time seems to have slowed, allowing them to fully absorb the importance of what they've created.

Maya eventually breaks the silence, her voice soft but clear. "Jaxon, I think we've found something special here. It's not just about the food–it's about everything we've been through, everything we're working towards. This dish, this moment–it's a reflection of our journey, and it feels like the start of something bigger."

Jaxon nods, his heart swelling with a mix of emotions. "I agree, Maya. What we've done today... it's more than just cooking. It's about healing, about finding our way through the darkness and coming out the other side. I think we're onto something that could really make a difference, not just for us, but for others too."

They continue to talk late into the afternoon, their conversation shifting from the personal to the practical as they discuss the possibilities for their future collaboration. Ideas flow freely, each one building on the last, as they brainstorm ways to bring their shared vision to life. By the time they part ways, the seeds of a new project have been planted, one that combines their talents and experiences in a way that feels both exciting and deeply meaningful.

As Maya leaves the studio, the sun dips below the horizon, leaving the room bathed in the soft glow of twilight. Jaxon stands alone for a moment, taking in the quiet stillness that has settled over the space. He feels a sense of peace that has eluded him for so long, a calm certainty that he is finally on the right path.

The studio, now empty, seems to hold the echoes of the day's events, the energy of their collaboration lingering in the air like a tangible presence. Jaxon knows that this is just the beginning, that there is still much work to be done, but for the first time in a long time, he feels truly hopeful about the future.

He takes one last look around the studio before turning off the lights and heading out the door. The city greets him with its usual hustle and bustle, but tonight, it feels different. The noise and chaos that once overwhelmed him now seem distant, almost irrelevant. Jaxon knows that he has found something more important, something that gives his life purpose and direction.

As he walks down the street, the cool night air clears his mind, allowing him to focus on the possibilities ahead. The journey has been long and difficult, but he is finally ready to embrace the challenges that lie ahead. With Maya by his side, he feels confident that they can create something truly extraordinary, something that will not only heal their own wounds but also help others find their way to the light.

The night sky stretches above him, vast and filled with stars, each one a reminder of the infinite possibilities that await. Jaxon smiles to himself, feeling a sense of contentment that he hasn't experienced in years. The future is uncertain, but for the first time, he feels ready to face it head-on, knowing that he is no longer alone.

The sun has dipped below the horizon, casting the studio in a warm, golden glow that complements the rustic charm of the space. Jaxon and Maya, hands synchronized in a dance of culinary precision, plate their creation with meticulous care. The air is filled with the aroma of fresh herbs, roasted vegetables, and the subtle hint of spices, each scent telling a story of resilience and renewal.

Natural light streams through the large windows, casting soft, twin shadows across the rustic countertops. Herbs and spices in crystalline jars catch the sunlight, their familiar scents mingling with the tantalizing aroma of their dish. Each movement, deliberate and assured, speaks of their newfound synergy–Jaxon's geometric precision intertwines seamlessly with Maya's fluid artistry. The brushed stainless steel of the kitchen island gleams, a silent witness to their creative fusion.

The dish before them, an intricate tapestry of colors and textures, stands ready. Jaxon places the final sprig of micro basil, a delicate punctuation to the symphony of flavors. Maya's touch adds a drizzle of saffron-infused reduction, casting a golden hue over the plate. They exchange a glance, a silent testament to the journey they've embarked on together.

Friends gather around the communal dining table, their faces eager and expectant. The table, a centerpiece of the studio, represents not just a place to eat but a nexus of shared experiences and communion. Wooden surfaces are rough-hewn, adorned with simple arrangements of wildflowers and candles flickering in glass jars, creating an atmosphere of warmth and intimacy.

Jaxon steps forward, holding the dish with steady hands. "What you see here is the result of a week of experimenting," he begins, his voice carrying both pride and humility. "We took what we've each learned, not just about cooking, but about ourselves and our paths. This dish is a reflection of that."

Maya picks up the narrative, her tone infused with enthusiasm. "We wanted to merge our techniques–Jaxon's precision and my improvisation–to create something unique. Each element here represents a piece of our journey." Her eyes sparkle with a deep warmth as she connects with each person around the table.

The group leans in as Jaxon and Maya place the dish at the center. The vibrant array of flavors calls to them, an edible testament to resilience and creativity. As forks delve in, murmurs of pleasure ripple through the room.

"Oh my goodness, the flavors!" exclaims one friend, her eyes widening in surprise. "It's like nothing I've ever tasted before." The mix of spices and textures unfolds, layer by layer, each bite a narrative unto itself.

Conversations ignite with fervor, stories of past culinary mishaps and triumphs erupting in contagious laughter. Jaxon and Maya bask in the shared joy, the atmosphere thick with camaraderie. Plates clink softly, and the aroma of the dish mingles with the sounds of genuine connection.

Brett, a long-time friend of Jaxon, raises his glass. "To new beginnings, and to friendships old and new," he toasts, his eyes gleaming with pride and respect. The studio feels alive, a sanctuary of healing and hope where past pains give way to transformation.

As their friends savor the intricate flavors, heartfelt discussions blossom. Emotions unfold like lotus petals, revealing depths of understanding and appreciation. Through bites of food, friendships deepen, layers of personal history blending into a collective tapestry of support.

Laughter bubbles up, stories flow freely–childhood memories, mishaps, and moments of revelation. Each tale acts as a thread, weaving the group closer together. Jaxon and Maya, though physically apart in the room, are synchronized in spirit. Their collaboration has forged not just a dish, but a community.

The night deepens, the sky outside darkening into a rich indigo, while inside, the studio is a haven of warmth and light. The candles on the table flicker softly, casting dancing shadows that reflect the animated conversations and laughter filling the space. The meal they've shared has become more than just a culinary experience– it's a bonding ritual, one that strengthens the connections between everyone present.

As the evening wanes, the room settles into a contented hum. Firelight dances across smiling faces, and the studio now glows with the assurance of bonds forged through shared creativity. Their dish, now an empty platter, stands as evidence of their journey and the promise of future endeavors.

In a quiet, deliberate moment, Jaxon and Maya's eyes meet across the table. There's no need for words; the shared glance encapsulates their pride and the unique magic they have created. The satisfaction of their work resonates deeply, echoing a silent commitment to continue inspiring and growing together.

Their friends, filled with both surprise and joy, linger, reluctant to leave the shared sanctuary. The culinary studio, now a crucible of healing and artistic expression, stands ready to host more moments of connection and transformation.

As the last of the friends depart, their departing laughter still echoing in the candlelit room, the promise of enduring relationships hangs palpable in the air. Jaxon and Maya stand together, side by side, souls intertwined through creativity and shared purpose.

The warmth of the evening lingers in the studio, even as the last of their friends depart. The remnants of the evening's feast are scattered across the long wooden table–empty plates, half-drunk glasses of wine, and a few crumbs that tell the story of a meal well enjoyed. The soft glow of the candles continues to flicker, casting a golden hue over the space, making it feel both intimate and expansive.

After the guests leave, Jaxon and Maya turn toward the kitchen, which lies in the aftermath of the day's chaos. As they begin cleaning up, laughter punctuates the clinking of dishes and the swishing of sponges. The studio, bathed in the golden hues of the setting sun, feels imbued with a sense of achieved harmony. Jaxon wipes down the countertops, each stroke carrying the weight of today's intricate culinary dance. His eyes occasionally meet Maya's, and both share silent, knowing smiles.

Maya's movements are fluid and precise, her laughter light and unguarded. As she stacks freshly washed platters, she recalls the first time she attempted to cook a complex dish, a story layered with nostalgic humor. Jaxon chuckles, the sound a comforting tremor in the studio's now tranquil atmosphere. With each anecdote, the tension of earlier demands dissolve, leaving only the warmth of shared company.

As they continue their cleaning ritual, the conversation flows naturally, shifting from lighthearted banter to deeper reflections on the day's events. Jaxon finds himself reflecting on the surprising synergy of their collaboration. He sees Maya as more than an ally, more than a rival. The culinary studio, with its inviting earthy tones and natural light, stands as a testament to his rebirth, an arena where he can reconstruct his scattered past into a meaningful present.

Memories surface–of the frenetic energy of kitchens past, the perfectionist pursuit of culinary artistry, and the destructive allure of addiction that once shattered his world. Each dish he created then was a battle between his talents and his demons, a conflict that left scars now both literal and figurative. Yet, today felt different. Today, the act of creating food became a symphony rather than a struggle, in large part due to Maya's influence.

In her own silent moments, Maya reflects on her journey. Losing her friend to addiction had driven her relentless pursuit of understanding, a bittersweet impetus that melds resilience with vulnerability. She thinks of how working with Jaxon has reignited her belief that creativity and empathy could transform pain into art. Their shared pasts–his filled with the chaos of addiction, hers marked by loss–forge an unspoken bond.

Eventually, they settle into chairs with steaming cups of herbal tea, the fragrant steam curling into the quiet air. Jaxon takes a sip, savoring the warmth before speaking. "Maya, today was... it was more than I expected. You bring something to the table that I can't quite define, but it's healing. Your presence here has been a balm."

Maya looks at him, her eyes reflecting the depths of unspoken thoughts. "Jaxon, you've brought out something in me today too. It's like our stories, our scars, they somehow blend into the flavors we create. I never thought cooking could be this cathartic."

They share a contemplative silence, sipping their tea. The studio, once filled with the clamor of cooking, now breathes with peaceful energy. The aromas of the day still linger, a sensory reminder of their creative efforts.

Jaxon sets down his cup, his gaze thoughtful. "I've been thinking about how we can keep this going. Our collaboration... it feels like more than just cooking."

Maya nods, the hint of a smile playing on her lips. "I've been pondering the same thing. What if we create a series of dishes based on our shared experiences? Something that tells a story through taste–our story, and the stories of those we want to help."

Jaxon's heart swells with a mix of excitement and calm determination. He sees the potential for their partnership extending beyond the confines of the kitchen studio–a fusion of past and present, chaos and creativity. "That sounds like a way to not just cook, but to continue healing, to grow."

They raise their teacups in a silent toast, the soft clink of porcelain heralding their unspoken promises. As the evening fades into night, the culinary studio stands as a sanctuary of possibilities, a space where trauma transforms into trust, and individual shadows blend into a shared light.

With the day's work complete and the studio restored to its serene state, Jaxon and Maya take a moment to appreciate the quiet stillness that has settled over the space. The night outside is calm, the stars twinkling faintly in the distance, as if watching over them.

As they stand by the windows, looking out at the city lights that stretch into the horizon, Jaxon feels a deep sense of contentment. The future, once uncertain and filled with shadows, now feels open and full of potential. With Maya by his side, he knows that they can create something truly meaningful, something that transcends the boundaries of the kitchen and touches the lives of others in profound ways.

They continue to talk, their voices soft and filled with the kind of warmth that only comes from shared understanding. The bond between them, forged through their collaboration and their shared histories, feels stronger than ever. They discuss their plans for the future, the projects they want to undertake, and the impact they hope to have on the world.

As the night wears on, the studio becomes a haven of creativity and connection, a place where dreams are nurtured and brought to life. Jaxon and Maya, united by their shared vision, are ready to take on whatever challenges lie ahead, knowing that they have each other to rely on.

The final hours of the night are spent in quiet companionship, the kind that requires no words, only the comfort of being in the presence of someone who truly understands. When they finally part ways, the promise of the future lingers in the air, like the faint scent of spices that still clings to their clothes.

Jaxon watches as Maya disappears into the night, a smile on his lips and a sense of hope in his heart. The journey they've embarked on together is just beginning, and he knows that it will be filled with challenges, but also with joy and discovery.

The studio, now empty and quiet, feels like a sacred space, a place where something truly extraordinary has been set in motion. As Jaxon turns off the lights and locks the door, he knows that he will return tomorrow, ready to continue the work they've started, ready to keep building on the foundation they've laid.

The night is cool and still as Jaxon walks home, his mind filled with thoughts of the future. He feels a deep sense of gratitude for the path that has brought him to this point, for the people who have supported him along the way, and for the opportunities that lie ahead.

As he reaches his front door, Jaxon pauses for a moment, looking up at the night sky. The stars seem to shine a little brighter tonight, as if they too are celebrating the new beginning that has just been set in motion. With a final smile, Jaxon steps inside, ready to embrace whatever the future holds.

Chapter 19

THE ALCHEMY OF THE SELF

Evening light filters through the high windows of Jaxon's culinary studio, casting a warm glow upon the polished marble counters and gleaming kitchen appliances. The air is thick with the heady scent of spices—saffron, cumin, and cardamom—a sensory prelude to the gathering. Jaxon moves methodically, placing bowls of fresh produce, herbs, and neatly arranged utensils within easy reach. Every detail reflects his meticulous nature, a vestige from his past life as a renowned chef, now serving as the foundation for his quest to blend culinary artistry with spiritual enlightenment.

Jaxon's thoughts swirl like the fractal patterns he continually sees. He recalls his journey from addiction to his awakening, the lightning bolt that realigned his existence. This studio stands as a symbol of his transformation, a sanctuary where the alchemy of food meets the transcendence of the soul. The kitchen, once a battlefield of intense culinary competition, has become a sacred space for introspection and connection. Each ingredient, each tool, holds a potential to unlock deeper truths, mirroring the intricate steps of his inner journey.

He walks to the center of the studio, his fingers brushing lightly against the smooth surface of the kitchen island. His mind races with the possibilities that tonight's gathering could bring. He knows that the seekers who will soon fill this space are not just here to cook; they are here to explore the intersections of food, spirituality, and self-discovery.

The kitchen radiates readiness, every detail carefully arranged, from the hand-carved wooden spoons to the artisanal bowls filled with ingredients sourced from local farms. This is not just a culinary studio—it is a crucible of transformation.

The first guest arrives—a slender woman with flowing auburn hair and a serene demeanor. She introduces herself as Lila, a yoga instructor who believes in the harmony of body and mind through mindful eating. Her eyes sparkle with curiosity as she shares her personal quest for balance, each word carrying the weight of her spiritual journey. Jaxon receives her story with a nod, recognizing the echoes of his own struggles in her narrative.

As Lila unpacks her thoughts, she describes how food has always been a crucial part of her practice. "Every meal is an opportunity to center myself," she explains, her voice soft but firm. "It's not just about nourishing the body; it's about connecting with the earth, with the energy that flows through all living things."

Jaxon listens intently, his mind already weaving connections between her words and his own experiences. He can see that Lila's approach to food is rooted in a deep understanding of the interconnectedness of all things, much like his own philosophy.

Next, a tall man with salt-and-pepper hair steps in, his presence commanding yet gentle. He is Samuel, a retired marine biologist who found solace in the mystical properties of seaweed, viewing it as a bridge between marine life and human consciousness. As he unpacks his tale, the smell of ocean brine seems to infuse the room, blending seamlessly with the other aromas. Jaxon feels a kinship with Samuel's search for meaning in natural elements, understanding the profound connections between the earth and the human spirit.

Samuel's voice is rich and resonant as he speaks of his discoveries beneath the waves. "There's something almost magical about the way seaweed grows," he says, his eyes distant as if recalling the depths of the ocean. "It's like the ocean's own version of the fractal patterns you've described, Jaxon. Each strand is a microcosm, a reflection of the greater whole."

Jaxon nods, feeling a deep sense of connection with Samuel's insights. He knows that the sea, like the earth, holds secrets that are closely intertwined with the mysteries of the human soul.

Maya arrives last, her energy a mix of excitement and apprehension. The unspoken tension between her and Jaxon surfaces as they exchange glances–two seekers on parallel yet divergent paths. Her attire, a blend of bohemian elegance and intellectual flair, mirrors her complex persona. Maya recounts her adventures in Mexico, where she dived into ancient culinary practices and the spiritual significance of each meal. Her experiences are imbued with the vibrant colors and spices of her heritage, bringing a sensory tapestry into the space.

She speaks with an enthusiasm that is almost infectious, her hands moving animatedly as she describes the rich culinary traditions she encountered. "In Mexico, food is more than just sustenance," she says, her eyes shining. "It's a spiritual experience, a way of connecting with the ancestors, with the earth, and with each other. Every dish tells a story, every spice carries a piece of history."

Jaxon can't help but feel a mix of admiration and trepidation as he listens to Maya. Her passion for food and spirituality is palpable, but he also senses the underlying tension between them, a tension that has yet to be fully resolved.

As the last of the guests arrives, Jaxon's voice gently calls everyone to gather, initiating the evening with a grounding meditation. They form a circle, sitting on plush cushions around a low wooden table laden with earth-toned textiles. The soft hum of a Tibetan singing bowl fills the room, its vibrations harmonizing with the aromatic symphony. Jaxon's words guide them inward, fostering an atmosphere of openness and trust. He reflects on the significance of their gathering, the convergence of varied paths seeking a united purpose.

He begins by asking everyone to close their eyes and focus on their breath. The room falls into a deep, meditative silence, broken only by the gentle sound of the singing bowl. Jaxon feels the energy in the room shift as each person begins to relax, their minds opening to the possibilities of the evening.

As they breathe together, Jaxon's own thoughts begin to quiet. He contemplates his role not just as a facilitator, but as a guardian of the profound knowledge he has accessed. Doubt lingers, a shadow cast by his tumultuous past. Can he truly guide these seekers without faltering? The pressure of their expectations weighs on him, yet he finds solace in the shared breath, the collective intention to explore and understand.

As the meditation concludes, the group begins to share their experiences. Lila speaks first, detailing her relationship with food as a spiritual practice. Her voice is soft, yet it carries the strength of her conviction. She describes how every bite becomes a ritual, a way to ground herself in the present moment.

"I've found that the way I eat reflects the way I live," Lila says, her voice filled with a quiet reverence. "When I eat mindfully, I'm more present in every aspect of my life. It's like each meal becomes a prayer, a way of giving thanks for the abundance of the earth."

Samuel follows, discussing how his encounters with marine life have transformed his perspective on interconnectedness. He describes the fractal beauty of coral reefs, their similarities to the neural pathways of the human mind.

"The ocean has taught me so much about the interconnectedness of all things," Samuel says, his voice rich with experience. "The way that life in the ocean is so intricately connected—it's a reminder that we, too, are part of a much larger whole. When I think of food, I think of it as a way to honor that connection, to nourish not just our bodies, but our souls."

The dialogue weaves a tapestry of narratives, each thread contributing to a richer understanding. Jaxon listens intently, his gentle inquiries drawing deeper reflections from the group. Though he remains a silent observer for most of the conversation, his presence is a grounding force, binding their stories into a coherent whole.

Maya's turn comes, and her words pour forth with a fervent intensity. She speaks of her philosophical quests, the sleepless nights spent pondering the cosmic significance of everyday actions. She shares her insights on how food, prepared with intention, can be a doorway to higher consciousness. Her stories resonate with

Jaxon, stirring memories of his earlier struggles and the divine revelations he uncovered through hardship.

"Food is more than just sustenance," Maya says, her voice filled with passion. "It's a way of connecting with the divine, of experiencing the sacred in the everyday. When we cook and eat with intention, we're not just nourishing our bodies—we're feeding our souls, connecting with something greater than ourselves."

The room hums with shared understanding, each seeker contributing to an atmosphere of profound connection. Aromas of cinnamon and roasted vegetables mingle with the subtle murmur of voices, creating an ambiance both comforting and inspiring. These scents, these tastes, these stories—they are all facets of the same infinite fractal, each reflecting the complexity and beauty of their collective journey.

Jaxon takes a deep breath, feeling the weight of the evening settle over him. He knows that tonight is not just about cooking—it's about creating a space where these seekers can explore the deeper connections between food, spirituality, and self-discovery.

As the conversation continues, Jaxon begins to feel a sense of calm settle over him. The doubt that had been gnawing at him earlier begins to dissipate, replaced by a quiet confidence. He realizes that he doesn't have to have all the answers; he just needs to be present, to hold space for these seekers as they embark on their own journeys of discovery.

The evening progresses, the light outside fading as the warmth of the studio grows. The group continues to share their stories, each one adding a new layer of depth to the conversation. Jaxon listens, his heart swelling with a sense of gratitude for the opportunity to be part of this shared experience.

As the night deepens, the conversation begins to wind down, the seekers settling into a comfortable silence. Jaxon looks around the room, feeling a deep sense of connection with each person present. He knows that tonight is just the beginning of a journey that will continue to unfold in the days and weeks to come.

With a final deep breath, Jaxon brings the evening to a close. He thanks the group for their openness and vulnerability, for the courage it took to share their stories. As the seekers begin to gather their things, Jaxon feels a sense of peace settle over him. He knows that he has found his purpose, that this studio will continue to be a sanctuary for those seeking to explore the deeper connections between food, spirituality, and self-discovery.

The last of the seekers depart, leaving Jaxon alone in the quiet studio. He stands in the center of the room, taking in the warmth and light that still linger. He feels a deep sense of fulfillment, knowing that he has created a space where people can come together to explore the deeper aspects of their lives.

As he begins to clean up, Jaxon's thoughts turn to the future. He knows that there is still much work to be done, but he is ready for the challenge. He feels a renewed sense of purpose, a quiet determination to continue on this path, to continue guiding others on their own journeys of discovery.

The night is cool and still as Jaxon locks up the studio and heads home. The stars twinkle faintly overhead, a reminder of the vastness of the universe and the infinite possibilities that lie ahead. With a final glance at the night sky, Jaxon steps inside, ready to embrace whatever the future holds.

As the aromas of cardamom and saffron mingle with the lingering scents of rosemary and thyme, the cozy roundtable in Jaxon's culinary studio hums with an electric energy of anticipation. The soft lighting flickers like stardust, intertwining with the warm, reflective atmosphere that Jaxon has meticulously crafted. Plates of leftovers from earlier creations–a symphony of vibrant colors and intricate flavors–sit in the center, inviting discussion and introspection.

The seekers gather around the table, their faces lit by the warm glow of candlelight. There's a sense of expectancy in the air, a feeling that something profound is about to unfold. Jaxon senses it too, the energy in the room vibrating with the potential for deep revelation.

A seeker, her eyes gleaming with curiosity, speaks up, her voice lilting with a mix of awe and trepidation. "Jaxon, can you tell us what enlightenment truly means? How does one embark on the path to achieve it?"

Jaxon leans forward, the flicker of the candlelight catching the edges of the scar on his cheek, a reminder of past trials and growth. He clasps his hands together thoughtfully before responding, the soft timbre of his voice imbued with a quiet intensity. "Enlightenment, to me, is not a destination but a journey–a continual unfolding of understanding, much like the fractal patterns I observe in food and memory. Each experience shapes us, refracting shards of clarity that guide us forward."

The seekers, each engaged in their own dance of contemplation, nod in varying degrees of agreement. Dr. Miles Carter, always the skeptic, tilts forward, his wire-rimmed glasses catching the light. "But Jaxon, can enlightenment be something we teach? Or is it an endeavor each person must undertake on their own?"

Maya Santos, seated across from Jaxon, her eyes sharper than the blade of a chef's knife, joins the conversation. "Preparation of food provides a profound metaphor for growth. Without the right ingredients and care, a dish falls flat. We can guide each other, offer our experiences as seasoning, but the essence of enlightenment must come from within."

Her words are like a glint of light diffusing through a prism, complementing Jaxon's musings while subtly asserting her own presence in the room. Allegra Frost, her eyes like emerald orbs of mystic depth, gently interjects, "The fractal patterns Jaxon speaks of–these are sacred. They mirror the interconnectedness of our journeys. We are all threads in this vast tapestry."

Jaxon's heart swells with an unspoken recognition that here, in this moment, the culinary studio transforms into a cauldron of collective wisdom. He draws a deep breath, the scent of rosemary grounding him. "Every dish I prepare, every ingredient I choose, mirrors the chaos and order within our lives. Enlightenment is the continuous interplay between these states. As we grow, we don't discard the darkness; we fold it into the light."

A palpable wave of introspection sweeps the room. The seekers digest his words alongside the remnants of his culinary expertise. Dr. Miles's analytical mindset refuses to settle easily. "Enlightenment as a personal journey resonates, yet, can it not be accelerated by shared knowledge and structured learning?"

Jaxon smiles, acknowledging the merit in Miles's question. "Yes, shared knowledge defends against stagnation. However, it's not a shortcut, but a catalyst. The equation behind our consciousness must be experienced beyond mere comprehension."

A murmur of agreement ripples through the group, carrying with it the taste of cardamom and the warmth of freshly-baked bread. The candlelight dances upon their faces, a visual testament to the inner truths they each wrestle with.

Maya then counters, her voice imbued with both passion and reverence, "Jaxon speaks to the core—personal experience shapes the enlightenment pursuit. Yet, is the path to understanding less valid without communal steps?"

Her competitiveness with Jaxon is apparent, a constant undercurrent in their interactions that now bubbles to the surface. Allegra places her hand gently on the table, her tattooed spirals reflecting sacred geometry, mystical and precise. "Community offers balance, a mirror reflecting our truths. But like an intricate recipe, it's the understanding of each element that leads to mastery."

The debate becomes more animated, each seeker contributing ingredients to this philosophical potluck. Laughter mingles with earnest discourse, creating a nourishing broth of shared insights and divergent viewpoints. Emotions ride high; hands gesticulate, eyes flash with conviction and doubt.

Jaxon observes the ebb and flow of the conversation, his mind weaving connections between the seekers' words and the fractal patterns that have become such a central part of his understanding. He sees the debate not as a conflict, but as a necessary tension, a way for each seeker to refine their own understanding of enlightenment.

As the evening deepens, the conversation reaches a crescendo of unresolved questions and burgeoning insights. The seekers glance at each other, their faces glowing, not just with candlelight but with the ferment of ideas and camaraderie.

Jaxon lets the discussion ebb, allowing a peaceful silence to fill the studio. The flickering lights, the aromatic spices, the living snapshot of unity all coalesce into a moment of collective reflection. The debate may remain unresolved, but an unspoken understanding weaves through them, leaving each to ponder the depths of enlightenment on their unique paths.

As the seekers begin to settle into the quiet, Jaxon feels a sense of peace wash over him. He knows that the questions they've been grappling with tonight are not ones that can be answered easily–or perhaps at all. But he also knows that the process of asking these questions, of wrestling with the unknown, is itself a form of enlightenment.

He takes a deep breath, savoring the quiet that has descended over the studio. The conversation may be over for now, but the journey continues, each seeker carrying the evening's insights with them as they move forward on their paths.

Jaxon stands at the rustic kitchen island, his hands deftly arranging a spectrum of spices, their colors vivid against the polished marble surface. The culinary studio hums with a quiet energy, a sanctuary of warmth and light amidst the fading evening. Aromas of cumin, coriander, and cardamom swirl through the air, each scent tugging at memories of his former life–a life steeped in addiction and the relentless pursuit of culinary perfection.

The seekers drift into the studio, each one a pilgrim on their path to understanding. They share stories as they settle, their voices a gentle murmur against the tranquil backdrop. Jaxon observes them with a careful eye, detecting the subtle signs of their journeys–the weariness in their eyes, the hopeful lilt in their voices. Each seeker carries their own burdens, their struggles etched into their very being.

As the last guest finds a seat, Jaxon steps forward, his presence commanding yet comforting. He initiates the evening with a grounding meditation, guiding them through the intricate dance of breath and awareness. The room falls silent, filled only with the rhythmic cadence of collective inhalation and exhalation. He feels the energy shift, a palpable sense of openness settling over the group.

The silence is thick, not with tension, but with the weight of collective introspection. Jaxon senses the readiness in the room, the unspoken commitment of each seeker to the journey they've embarked upon. His own breath synchronizes with the group, creating a rhythm that feels almost like a heartbeat, pulsing through the room.

With the meditation complete, the seekers begin to share their experiences. One by one, they recount their encounters with food, consciousness, and spirituality, each narrative a thread weaving into the larger tapestry of the evening. Jaxon listens, his mind a conduit of empathy and understanding. He recognizes fragments of his own story in theirs—the relentless quest for enlightenment, the intersections of joy and despair.

There's a young man with a troubled past who speaks of his battle with addiction, his voice trembling as he describes the nights spent in darkness, searching for a way out. "I found solace in cooking," he says, his voice barely above a whisper. "It was the only thing that made sense to me, the only thing that gave me a sense of control."

Jaxon listens, his heart heavy with empathy. He knows all too well the struggle of addiction, the way it can consume your life, leaving you feeling like a shell of your former self. But he also knows the power of food, the way it can heal, the way it can bring you back to yourself.

When it's finally his turn to speak, Jaxon takes a deep breath, feeling the gaze of the seekers upon him. He begins to share his story, his voice carrying the weight of years spent in the throes of addiction. He speaks of the numbing haze of drugs and alcohol, the moments of utter hopelessness. But he also recalls the fleeting sparks of inspiration, the whispers of a deeper truth that kept him tethered to his essence.

"The lightning strike was my catalyst," Jaxon says, his eyes distant as he revisits the pivotal moment. "In that instant, I was realigned with the universe, my mind a canvas for a divine equation that transcended the mundane. It was as if I had been given a second chance to rewrite my existence."

The seekers listen, enraptured by his tale. One of them, a woman with eyes that shimmer like the ocean, speaks up. She shares her own struggles, the battles with depression and the slow climb towards understanding. Her voice wavers, but there's a quiet strength in her words, a resonance that vibrates through the room.

"I've spent years feeling lost," she says, her voice trembling slightly. "But I've found that when I cook, I feel a sense of purpose, a connection to something greater than myself. It's like the act of creating food brings me back to life, even on the darkest days."

The discussion broadens, as other participants begin to open up about their own journeys. Darkness and light intermingle in their stories, each recounting moments of despair countered by the pursuit of knowledge and the quest for balance. There is a shared recognition of the necessity to embrace inner chaos, a theme that mirrors the fractal patterns Jaxon now perceives in the universe.

He listens, feeling a profound sense of connection with the group. Their words echo the insights he's gained through his exploration of altered states and the cosmic dimensions that lie beyond human cognition. "Embracing our inner chaos is not just crucial," he says, his voice steady. "It's essential for growth. It's through this chaos that we find the patterns, the fractals that guide us towards a deeper understanding."

As the conversation flows, the group dynamics shift, their connections deepening with every shared story. There's a subtle comfort in their vulnerability, an unspoken recognition of shared humanity. The seekers nod and murmur in agreement, the atmosphere thick with empathy and support.

The meal that follows becomes more than just sustenance. It's a communion, each dish a symbolic offering of their collective journey. Jaxon watches as the seekers pass plates and bowls, their interactions gentle and attentive. The aroma of spices mingles with the hum of conversation, creating a tapestry of sensory and emotional nourishment.

As they eat, Jaxon reflects on his own transformation, his mind oscillating between the past and the present. He recalls the depths of his addiction, the nights he spent lost in a haze, and how those experiences carved pathways in his mind that led him to this very moment. Every bite of food he prepares now is infused with meaning, a reflection of the fractal equation that binds him to the seekers and the universe.

The seekers, too, engage in silent reflection as they savor the meal. Their faces soften with contentment, their eyes alight with newfound insights. Each one of them can feel the weight of the evening's discussions, the profound sense of unity that has been cultivated in this sacred space.

In the warm glow of the studio, the shared meal becomes a symbol of their enriched connection, a testament to the power of shared wisdom and collective journeys. Jaxon feels a swell of gratitude, not just for his own awakening but for the opportunity to guide others on their paths. As the evening draws to a close, he knows that the bonds forged tonight will continue to resonate, their collective enlightenment a beacon for the journeys yet to come.

Soft ambient light bathes the culinary studio, creating a cozy cocoon of warm hues that embrace the walls filled with spices and cooking utensils. As night settles deeper, it's a sanctuary of reflection and transformation. The seekers, a diverse group of individuals embodying as many stories as aspirations, form a tight circle. Each face is illuminated by the gentle flicker of candlelight, eyes reflecting both curiosity and a sense of belonging.

Jaxon, sitting at the end of the table, looks around, his gaze settling on each of the seekers. His heart swells with a mixture of pride and humble gratitude. He remains quiet, absorbing the palpable energy of the room. He notices the nervous flick of fingers, the contemplative expressions, and the subtle nods of encouragement exchanged among the guests. In this shared silence, the room hums with the unspoken promise of revelations.

The first to speak is a young woman named Elena. Her voice trembles slightly as she begins, "My intention for the future is to heal through food. Growing up, meals were the only moments of peace in our chaotic household." Her words carry a weight, each syllable a thread weaving a tapestry of her past. "I want to open a community kitchen where anyone can find solace and nourishment —not just of the body, but of the soul."

Jaxon leans back in his chair, his fingertips brushing against the smooth marble of the kitchen island. He contemplates the power of Elena's vision, seeing reflections of his own struggles and triumphs in her courage. He murmurs an acknowledgment, "Healing through food is a profound gift, Elena. It's a way to connect deeply, to mend what words often fail to touch."

Next is Marcus, a man who wears his years of searching like a rugged cloak. His voice is gravelly and carries the weight of many lifetimes. "I intend to find a balance between the spiritual and the physical. For too long, I've wandered in extremes, either too rooted in the material world or lost in the esoteric." He looks directly at Jaxon, a silent plea for understanding in his eyes. "I want to use culinary arts to bridge that gap."

Jaxon nods slowly, feeling the significance of Marcus's journey resonate in his own chest. "Balance is essential," he responds, his voice carrying the quiet strength of someone who has walked a thousand miles in search of equilibrium. "Your intention reminds us that every dish, every ingredient can be a step toward reconciling those inner conflicts."

As the sharing continues, the room becomes a canvas painted with their stories and aspirations. Each seeker peels back layers of their soul, revealing scars and dreams entangled in their pursuit of enlightenment. The atmosphere grows dense with an energy that is both grounding and elevating, a testament to the myriad paths converging in this space.

Jaxon's thoughts meander through the tales like a river through a valley, reflecting on his role in this sacred convergence. His mind drifts back to his past, the relentless grip of addiction, and the searing revelation brought by the lightning strike that changed everything. He sees fragments of his journey mirrored in their faces, a constellation of experiences interwoven with the same fractal patterns that now define his understanding of reality. This culinary studio, once a prison of his addictions, has transformed into a crucible of collective enlightenment.

One of the seekers, an older woman named Claire, speaks next. Her eyes glisten with unshed tears as she shares, "I intend to embrace the chaos within me and transform it into clarity. I have spent too long running from my shadow." Her voice tremors but holds a steadfast determination. "Through cooking and meditation, I want to discover the harmony within my own discord."

Jaxon feels a pang of empathy and recognition. Her words strike a chord deep within him–a melody of shared pain and triumph. "Embracing chaos is indeed a journey of immense courage," he says, his voice soothing yet authoritative. "It is through cooking, through the deliberate act of creating, that you can channel that chaos into something beautiful and meaningful."

As the final seeker finishes speaking, Jaxon feels the weight of the evening settling into a profound stillness. He stands, a bundle covered by a large linen cloth in hand. The room grows hushed as the seekers watch him unveil a dish that embodies their collective intentions. It's a vibrant tapestry of colors and textures–the intricate dance of spices and herbs creating a symphony for the senses.

"This dish represents our shared journey," Jaxon begins, his voice filled with reverence. "Each flavor is a reflection of the diverse paths we've walked and the unity we've cultivated here tonight." He distributes small portions to each seeker, the simple act of sharing imbued with ritualistic significance.

The moment of silent reflection that follows is almost tangible. Eyes closed, breaths synchronized, the seekers are united in a meditative embrace of the moment's sacredness. Jaxon feels the ripples of their shared consciousness, each one contributing to the larger wave of understanding sweeping over them. It's a silent vow to honor their intentions, to carry the wisdom of this night into the world beyond this sanctuary.

As the soft light continues to dim, the seekers depart in gentle waves, leaving the culinary studio with a deeper sense of interconnectedness and purpose. The remnants of their meal linger as a testament to the evening's journey—a mosaic of flavors, stories, and aspirations that will continue to nourish both body and soul.

Jaxon remains, standing in the soft glow of the now-quiet studio. He feels the threads of their shared experiences knitting together into a tapestry of profound unity and understanding. He knows that this night marks not just the end of a gathering, but the beginning of countless new journeys towards enlightenment. The studio, now a dimly lit sanctuary, hums with the energy of those who passed through it, leaving traces of their stories behind.

As he begins to clean up, Jaxon reflects on the evening, the diverse paths that converged in his studio. He feels a deep sense of fulfillment, knowing that this space will continue to be a place of transformation, where seekers can explore the deeper connections between food, spirituality, and self-discovery.

The night is cool and still as Jaxon locks up the studio and heads home. The stars twinkle faintly overhead, a reminder of the vastness of the universe and the infinite possibilities that lie ahead. With a final glance at the night sky, Jaxon steps inside, ready to embrace whatever the future holds.

Chapter 20

A SYMPHONY OF SILENCE

The sun was just beginning to stretch its rays across the horizon, painting the sky with hues of pink and gold as Jaxon stepped into the culinary studio. The early morning light bathed the room in a warm glow, making the polished marble countertops gleam like liquid silver. Fresh ingredients were meticulously arranged, their vibrant colors popping against the backdrop of the sleek, modern kitchen. The room felt like a symphony of potential, each element poised to come together in a harmonious crescendo.

Jaxon paused at the doorway, breathing deeply, savoring the mix of aromas that filled the air: the sharp tang of citrus, the earthy richness of wild mushrooms, the sweet and spicy notes of basil and cinnamon. This was his sanctuary, a space where the boundaries between the culinary and the cosmic blurred, where he could channel the energy of the universe into something tangible. Today, the kitchen was more than just a place to prepare food—it was a stage where his final performance would unfold, a culmination of everything he had learned on his journey from addiction to enlightenment.

Moving with deliberate grace, Jaxon approached the marble island at the center of the room. His hands moved instinctively, gathering spices, arranging them in a pattern that only he could see. To him, these spices were like stars in a constellation, each one representing a different aspect of the universe. He reached for the coriander seeds, feeling the warmth and depth they would bring to the dish, like distant galaxies adding depth to the night sky. Next, he picked up the delicate saffron threads, their golden hue reminding him of the fragile fabric of time.

As he crushed the spices with a mortar and pestle, the rhythmic motion grounded him in the moment. Each movement was a silent invocation, a prayer to the flavors locked within. The scents began to mingle, creating an aromatic symphony that filled the room, and Jaxon closed his eyes, allowing himself to be swept up in the sensory experience.

His mind wandered as he worked, drifting back to memories of triumph and pain. He recalled the nights spent in the dark corners of despair, seeking solace in substances that only deepened his isolation. He remembered the applause of diners in Michelin-starred restaurants, the accolades that had once defined him. But those memories felt distant now, like shadows cast by a long-forgotten past. Today, his focus was different. Every choice he made, every movement, was imbued with a deeper significance.

He thought of how these flavors represented facets of existence—cumin for the warmth of the human spirit, rosemary for the clarity of thought, lemon zest for the zest of life itself. Each ingredient was a piece of a larger puzzle, a fragment of the cosmic equation he had been piecing together ever since that fateful lightning strike.

The kitchen came alive with the sound of sizzling oil as Jaxon began to cook. The aroma of caramelizing onions filled the air, their sweetness spreading out like the threads of his own story—interwoven, complex, and essential. He added garlic, thyme, and a splash of white wine, the flavors mingling and deepening as they cooked. The room was filled with the comforting sounds of culinary alchemy: the bubbling of sauces, the rhythmic chopping of vegetables, the soft hum of the stovetop.

As Jaxon worked, he envisioned the patterns and fractals that now shaped his understanding of the universe. Each dish he created was a reflection of this complexity, a harmony of flavors that echoed the intricate dance of the cosmos. He carefully arranged slices of heirloom tomatoes into swirling mandalas, each piece fitting perfectly into the next, a kaleidoscope of nature's geometry. He placed micro-greens atop a canvas of beet puree, creating a lush, verdant landscape—a symbol of life's perpetual renewal.

With each finished plate, a sense of satisfaction washed over him. He traced delicate lines of truffle reduction across the main dish, its earthy aroma hinting at the unseen depths of the soil, a connection to the very building blocks of life. Tiny blossoms of edible flowers dotted the compositions, like the first blooms of spring breaking through winter's hold. His hands moved with a fluid grace that felt almost otherworldly, as if time itself had folded around him, allowing past, present, and future to converge in this moment.

Stepping back, Jaxon admired his creations. Each dish sat like a precious gem on the pristine countertop, a fractal representation of his journey from the depths of addiction to the heights of cosmic understanding. He took a deep breath, feeling the anticipation for the evening ahead, where these dishes would become bridges to the infinite.

He thought of the guests who would soon arrive, each of them seekers in their own right, drawn into this ephemeral communion. His heart swelled with a mixture of fulfillment and expectancy. Tonight was more than just a meal—it was a chance to share the wisdom he had gained, to connect with others on a deeper level, and perhaps, for some, to offer a taste of the divine.

The dining area of Jaxon's culinary studio was transformed into a sanctuary of light and warmth, as soft, amber-toned bulbs cast a gentle glow over the elegantly set tables. Deep sapphire linens draped the surfaces, their rich color a contrast to the gleaming silverware and the delicate crystal glasses that reflected the flickering candlelight. The atmosphere was almost sacred, with the air thick with the anticipation of the evening to come.

Guests began to arrive, each person bringing their own energy into the space. Some entered quietly, taking in the scene with a sense of reverence, while others exchanged excited whispers, their eyes darting from the beautifully arranged tables to the open kitchen where the aromas of Jaxon's creations mingled in the air. There were familiar faces—old friends and colleagues from Jaxon's past life as a celebrated chef, along with new acquaintances who had been drawn to him through his recent work. The room buzzed with a mix of excitement, curiosity, and a hint of apprehension.

Jaxon stood at the threshold of the dining area, watching as his guests took their seats. His heart pounded with a mixture of nerves and pride. He recognized the faces of those who had witnessed both his rise to fame and his descent into addiction, and he felt the weight of their expectations. His gaze lingered on his estranged sister, Angelica, who met his eyes with a tentative smile. There was hope in her expression, along with a myriad of unspoken questions.

As the last of the guests settled in, a hush fell over the room. Jaxon took a deep breath, feeling the gravity of the moment. He moved with deliberate grace, presenting the first course–a delicate consommé, clear and golden, adorned with edible flowers that floated on the surface like stars in a night sky. As he set each bowl before his guests, he began to speak, his voice steady and resonant.

"This evening is not just about food," Jaxon said, his eyes sweeping over the room. "It's about connection. It's about understanding the patterns that shape our lives, the fractal nature of the universe that we are all a part of. Each dish you'll experience tonight is a reflection of that journey–a journey that has taken me from the depths of despair to a place of clarity and purpose."

The room was silent as Jaxon continued, the weight of his words sinking in. He spoke of his journey, from the heights of culinary fame to the depths of addiction, and the transformative power of the lightning strike that had changed everything. "This consommé," he explained, gesturing to the bowls before them, "represents the primordial state of the cosmos, where chaos slowly gave birth to order. It's a reflection of the process I've undergone, the way I've come to see the world through new eyes."

The guests lifted their spoons, taking their first taste. A ripple of reaction spread across the table–eyes widened, breaths were held, and the quiet murmur of conversation began to build once more. Jaxon watched as the room came alive, the food acting as a catalyst for deeper discussion.

"This flavor," one guest, a former culinary school friend, remarked, "it's like you're guiding us to understand something deeper, something beyond just the taste."

Allegra Frost, her emerald eyes glowing in the candlelight, shared a memory of Jaxon from their younger years. "You once told me the secret to transcendent flavor is in the balance, the harmony of extremes," she said, a wistful smile playing on her lips. "I see now that you were talking about more than just food."

Maya Santos, seated beside Jaxon, added her own thoughts. "These dishes are more than sustenance," she said, her voice filled with quiet reverence. "They're a reflection of the infinite, a reminder that we are all connected through the patterns of existence. Tonight, we're not just eating—we're experiencing the universe in every bite."

Jaxon listened intently, his heart swelling with pride and gratitude. The conversations flowed around him, interwoven with laughter, tears, and moments of profound silence. Each guest shared their own reflections, their voices soft but resonant with the gravity of their experiences. As the meal progressed through courses—a delicate ceviche, a rich risotto, a dessert that embodied the concept of transcendence—the room became a haven of connection and revelation.

By the time the final course was served, the group had transformed from a collection of individuals into a cohesive unit, bound by the experience of shared introspection and understanding. Jaxon felt a deep sense of fulfillment as he observed the scene before him—a tapestry of shared humanity, woven together by the threads of regret, hope, and transformation.

In this moment, Jaxon realized that his role had shifted. He was no longer just a chef; he was a guide, a cosmic alchemist using food to forge connections between the earthly and the divine. This meal, this gathering, was a reconciliation not just with his past but with the infinite possibilities of what lay ahead.

As the guests lingered over their final bites, Jaxon took a step back, allowing himself a moment to breathe. The evening was far from over, but already he felt a profound sense of peace, knowing that the journey they had embarked on together was just beginning. The room was filled with the hum of contented conversation, the soft clinking of glasses, and the occasional burst of laughter—a symphony of human connection that resonated deeply within him.

Jaxon's gaze drifted over the table, taking in the faces of those who had joined him on this journey. He saw in them reflections of himself, each person a facet of the cosmic puzzle he had sought to understand. The air was thick with the aroma of caramelizing sugar and roasting herbs, a fragrant reminder of the evening's offerings. The guests, once fragmented, had come together in a communion of souls, each dish a testament to the cosmic truths Jaxon had unearthed.

The night was still young, with promises of more stories to share, more reflections to unfold. As the conversations settled into a smoother rhythm, Jaxon felt the weight of his journey lift, replaced by an overwhelming sense of fulfillment. This was not just a meal; it was a communion of souls, a bridge to the infinite possibilities that lay ahead.

The soft, twinkling fairy lights strung across the patio added a touch of magic to the evening, their gentle glow casting a warm light over the group as they transitioned from the dining area to the outdoor space. The air was cool, the stars above twinkling like scattered diamonds against the velvet sky. The patio had been transformed into a sanctuary of calm and contemplation, with comfortable seating arranged in intimate clusters, inviting the guests to relax and continue their conversations.

Jaxon stood at the head of the table, looking out at his guests. The evening had already been a profound experience, but he knew that the real connection was just beginning. He took a deep breath, feeling the cool night air fill his lungs, grounding him in the present moment. "Thank you all for being here tonight," he began, his voice steady and warm. "This evening is not just about the food we've shared, but about the connections we've forged, the understanding we've gained."

The guests shifted in their seats, their attention fully captured by the sincerity in Jaxon's tone. He spoke of his journey from addiction to enlightenment, detailing how the fractal patterns of the universe had revealed themselves to him through his culinary creations. "Each dish you've tasted tonight is more than just food," he explained. "It's a reflection of the cosmic truths I've come to understand–an invitation to explore the infinite possibilities that reside within us all."

Maya, seated to Jaxon's right, felt a complex mix of admiration and self-doubt as she listened. When Jaxon finished, she took a moment to gather her thoughts before speaking. "Through Jaxon, I've learned that food can be a gateway to consciousness," she began, her voice tinged with a blend of gratitude and vulnerability. "Our shared experiences in the kitchen have been transformative. Each flavor, each texture–it all carries profound meaning."

Her eyes met Jaxon's, a silent acknowledgment of their intertwined journeys. "Yet, alongside this transformation, I grapple with my own fears and insecurities," she admitted. "But it is through these very struggles that I've come to understand the depth of our bond. Your guidance has been invaluable, Jaxon, even when it felt like we were rivals."

As Maya's words hung in the air, the other guests began to open up, their voices layering stories of personal insight and discovery. A young man with an intense gaze spoke of how the fractal equation Jaxon created had initially overwhelmed him, thrusting him into a state of confusion. "But it also forced me to confront parts of myself I'd long ignored," he said, his voice cracking with emotion. "In that chaos, I found clarity."

Another guest, an elderly woman with a serene smile, added her own reflections. "The equation is a mirror," she said softly. "It shows us who we truly are, warts and all. It has been a difficult journey, but one that has led me to a deeper understanding of my own consciousness."

The conversations flowed, each story a unique thread weaving into a tapestry of shared human experience. Jaxon watched, his heart swelling with a sense of fulfillment. He raised his glass, an invitation for the others to do the same. "To our journeys," he toasted, his voice steady and clear. "To the lessons we've learned and the connections we've forged."

The glasses clinked together, the sound a harmonious note resonating through the night. The act of raising their glasses was more than a celebration; it was a recognition of the paths they walked together, each step filled with the potential for enlightenment. Laughter bubbled through the group, light and carefree, as the warmth of unity enveloped them.

Underneath the joy, subtle currents of personal conflict and aspiration swirled, unnoticed by all but the keenest observers. Maya glanced at Jaxon, her competitive nature still flickering beneath her serene exterior, while Jaxon sensed the lingering weight of his past mistakes. Yet, in this shared moment, these tensions seemed inconsequential, dissolved by the camaraderie that bound them.

As they reflected on their collective experiences, the guests revealed small facets of their personalities–quirky habits, distinctive speech patterns, and heartfelt confessions–creating a rich mosaic of individuality within the group. Each character, through their interactions, added depth to the narrative, making the sense of unity all the more poignant.

The evening stretched ahead, a canvas of possibility painted with the colors of their shared enlightenment. Under the twinkling lights and the vast night sky, the group embraced the moment, feeling the profound connection that came from navigating the mysteries of existence together.

The patio had become a haven of light and warmth, the perfect setting for the culmination of an evening that had been as much about introspection as it had been about food. Jaxon stood before his guests, the soft glow of candlelight dancing across their faces. He could feel the energy in the air–an electric current of connection and understanding that hummed through the night.

He took a deep breath, gathering his thoughts before speaking, feeling the significance of this moment settle over him like a gentle weight. "Thank you all," Jaxon began, his voice steady but laden with emotion. "Thank you for being here, for sharing this journey with me." His eyes swept over the faces of those gathered around the table. Each of them had confronted their own darkness and emerged transformed, reflections of his own turbulent path. He felt a surge of gratitude, not just for their presence, but for their collective courage to explore the unknown.

"The dishes you've just experienced are more than food," he continued. "They are manifestations of the knowledge I've gleaned, the cosmic symphony that vibrates through every atom of our existence." He paused, allowing his words to sink in, to resonate with his audience.

Jaxon took a moment to recollect the moments of clarity that had struck him during meditation and culinary epiphanies, memories that merged into a kaleidoscope of colors and shapes, revealing the fractal nature of reality. He channeled these insights into his address, articulating the beauty and complexity of the universe through his art.

"Culinary art is a vessel, a way to translate the ineffable into something tangible. Each spice, each herb, every slice and swirl, they all tell a story about the universe's intricate dance." He gestured to the plates on the table, each one a canvas portraying patterns inspired by his multidimensional experiences. "Tonight, these dishes reflect the fractal variations I've come to understand– each bite a portal to a new aspect of reality."

As Jaxon delved deeper into his reflections, the room grew silent, the guests wholly captivated. He spoke of the profound moments of self-discovery, when time folded in on itself and light bent in ways that defied perception. His words painted vivid images in their minds, creating a tapestry of interconnectedness that transcended mere human understanding.

He noticed Maya, her eyes brimming with tears, nodding in silent agreement. Her expression mirrored his own journey, a shared path of introspection and revelation. The connection between them felt palpable, underscoring the transformation wrought by their interactions.

"Through cooking, I've discovered that food is not just sustenance; it's a bridge between the mundane and the divine," Jaxon said, his voice taking on a reverent tone. "It has the power to transform, to elevate consciousness and create bonds that transcend our material existence." He recalled the avalanche of flavors exploding on his tongue during his first successful dish post-awakening, each ingredient whispering secrets of the cosmos into his psyche.

He saw Dr. Miles Carter, usually so analytical, leaning forward, eyes wide with interest. This was a man who had spent his life in the tangibles of neuroscience, now visibly moved by the poetic intangibility of Jaxon's words. It was a testament to the depth of their shared exploration.

Finally, Jaxon invited his guests to join him in a celebration of their collective journey. With a flick of his wrist, music unfurled into the air, a melody of ethereal notes that wrapped around them like a soft embrace. He moved toward the patio, where fairy lights glimmered like constellations against the night sky.

"Let's dance," he said, a smile spreading across his face. The invitation was met with joyous acceptance. They rose, one by one, and followed him outside, where the air was cool and tinged with possibility.

Under the vast expanse of stars, the group began to sway in unison, their movements a beautiful choreography of unity and enlightenment. Laughter and conversation mingled with the music, creating a symphony of human connection. Jaxon felt the weight of his journey lift, replaced by an overwhelming sense of fulfillment and a humbling connection to every soul present.

In this moment of shared joy and understanding, the fractal patterns of their experiences converged, weaving an intricate web that bound them all. Each step they took, each beat they danced to, reflected the harmony they'd found within themselves and with each other.

As the night deepened, Jaxon looked around, his heart swelling with gratitude and peace. The cosmic truths he'd unearthed now resonated within these bonds, illuminating their path forward. The sense of closure and celebration lingered in the air, a testament to the transformative power of their shared journey.

The chapter drew to a gentle close, leaving behind a tantalizing promise of continued exploration and enlightenment. Under the canopy of stars, they danced, a living embodiment of the unity and joy they'd discovered, and in this shared rhythm, they found the true essence of their journey together.

EPILOGUE

The kitchen is quiet now, the last of the guests having departed, their laughter and voices fading into the night like the dying embers of a once-roaring fire. The air is thick with the lingering aromas of the evening's creations, a symphony of spices and herbs that speak of journeys taken and wisdom earned. Jaxon Keller stands alone, his hands resting on the worn wooden counter that has borne witness to his greatest triumphs and deepest failures.

The universe is vast, and its mysteries are boundless, but tonight, in the soft glow of the fading candles, Jaxon feels at peace. He has danced with the cosmos, tasted the infinite, and learned that the beauty of life lies not in the certainty of answers, but in the pursuit of questions. The fractal patterns that once seemed so elusive now flow through him with ease, guiding his hand as surely as the instincts of a seasoned chef.

He recalls the faces of those who shared this journey with him—seekers of truth, each bringing their own ingredients to the table, each contributing to the rich, complex tapestry they wove together. They came to him hungry for knowledge, for understanding, and he fed them in ways that transcended the physical. Through their shared experiences, they found not just nourishment, but connection—a deep, abiding sense of belonging in a universe that is both vast and intimate.

As he extinguishes the last candle, Jaxon reflects on the journey that has brought him to this moment. The storms he weathered, the battles he fought, and the revelations that changed him are now etched into the very fabric of his being. He understands now that the quest for perfection is an illusion, a mirage that distracts from the true essence of life.

For in the end, it is not the final dish that matters, but the act of creation itself—the process of transforming raw ingredients into something greater, something that nourishes not just the body, but the soul. And in this process, he has found his purpose, his place within the infinite fractal of existence.

Jaxon Keller knows that his journey is far from over. There are still mysteries to unravel, still flavors to discover, and still connections to forge. But for now, he is content to simply be–to exist in the moment, to savor the now, and to trust in the wisdom that has guided him this far.

The kitchen is quiet, but within the silence lies the promise of new beginnings. For in every end, there is a new beginning, and in every beginning, the seeds of endless possibilities.

And so, with a heart full of gratitude and a mind open to the infinite, Jaxon Keller steps into the night, ready to continue his journey into the unknown.

ABOUT THE AUTHOR

Lance Overman is an introspective author and culinary enthusiast with a passion for unraveling the complexities of existence through the lens of spirituality. With a background in electrical engineering, Lance merges a methodical approach with an imaginative spirit, crafting narratives that delve into the intersection of science, art, and the human experience. In The Devine Equation, Lance draws from his deep appreciation for the culinary arts and his fascination with the fractal nature of the universe to create a story that invites readers to explore the profound connections between taste, memory, and cosmic understanding. Lance's work is heavily influenced by his personal journey—a path marked by exploration, transformation, and a relentless pursuit of deeper meaning. His writing reflects a belief in the power of storytelling to inspire introspection and foster connections, whether through the vivid portrayal of a dish or the intricate weaving of a narrative. For Lance, writing is not just a creative outlet but a vehicle for exploring the unseen forces that shape our reality. His goal is to craft stories that resonate on multiple levels, offering readers a taste of the infinite possibilities that exist within the ordinary and the extraordinary alike. Through his work, Lance seeks to challenge, captivate, and ultimately leave a lasting impact, whether by uncovering the delicate balance of flavors in a dish or by probing the deeper mysteries of the human condition.

www.ingramcontent.com/pod-product-compliance
Lightning Source LLC
Chambersburg PA
CBHW052247220526
45471CB00001B/230